P9-CEZ-423

IF YOU CAN'T TRUST A MAN'S HANDSHAKE,

YOU CAN'T TRUST HIS SIGNATURE.

Warren A. Bechtel

I'M VERY PROUD OF OUR PEOPLE. I KNOW THEY CAN BUILD

ANYTHING, UNDER ANY CONDITIONS, ANYWHERE

ON THE FACE OF THE GLOBE.

Stephen D. Bechtel Sr.

IT'S NOT IMPORTANT THAT WE BE THE BIGGEST.

IT'S IMPORTANT THAT WE BE THE BEST.

Stephen D. Bechtel Jr.

GIVE BECHTEL PEOPLE A CHALLENGE, AND

THEY'LL MAKE IT HISTORY.

Riley P. Bechtel

BUILDING A CENTURY

BECHTEL
1898 – 1998

ANDREWS McMEEL
A Smallwood & Stewart Book

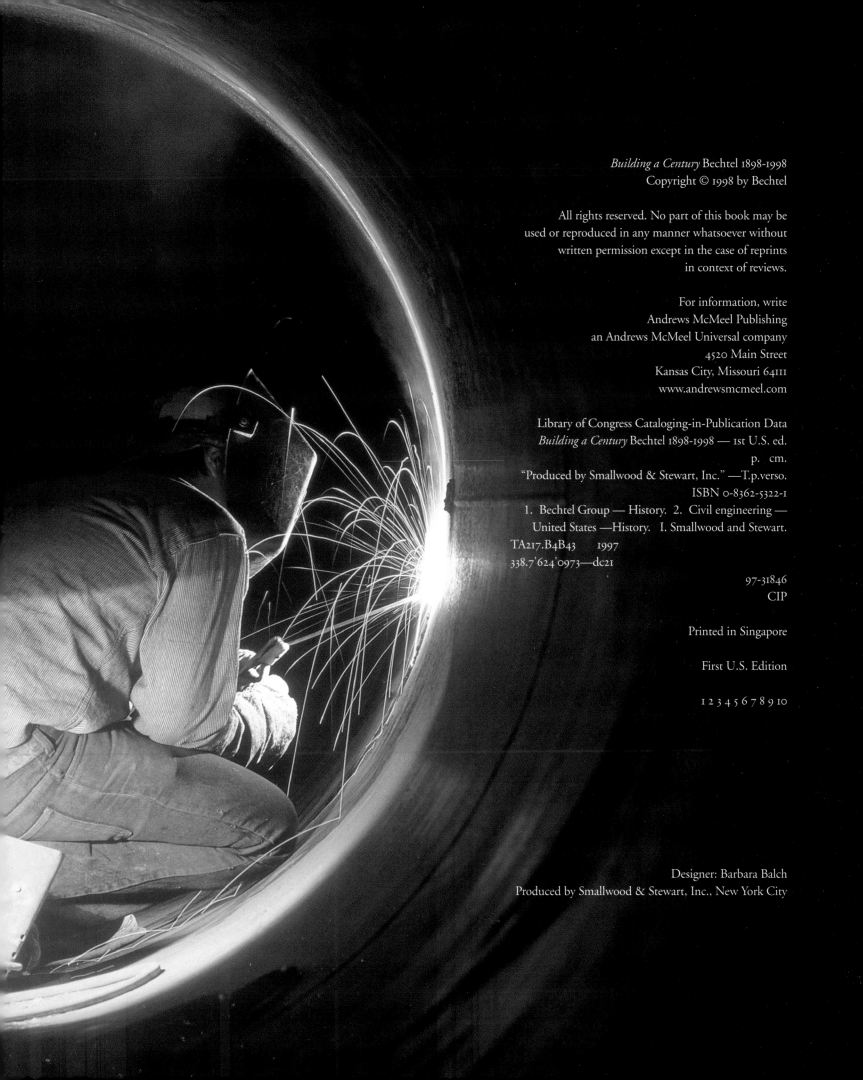

Building a Century Bechtel 1898-1998
Copyright © 1998 by Bechtel

All rights reserved. No part of this book may be
used or reproduced in any manner whatsoever without
written permission except in the case of reprints
in context of reviews.

For information, write
Andrews McMeel Publishing
an Andrews McMeel Universal company
4520 Main Street
Kansas City, Missouri 64111
www.andrewsmcmeel.com

Library of Congress Cataloging-in-Publication Data
Building a Century Bechtel 1898-1998 — 1st U.S. ed.
p.   cm.
"Produced by Smallwood & Stewart, Inc." —T.p.verso.
ISBN 0-8362-5322-1
1. Bechtel Group — History. 2. Civil engineering —
United States —History.   I. Smallwood and Stewart.
TA217.B4B43   1997
338.7'624'0973—dc21

97-31846
CIP

Printed in Singapore

First U.S. Edition

1 2 3 4 5 6 7 8 9 10

Designer: Barbara Balch
Produced by Smallwood & Stewart, Inc., New York City

CONTENTS

CHAPTER FIVE

A DECADE OF MEGAPROJECTS

CHAPTER SIX

GLOBAL DOWNTURN

CHAPTER SEVEN

BUILDING THE NEXT CENTURY

PART II : PROJECTS

FOREWORD

NESTLED AMID SAN FRANCISCO'S HIGH-RISES is an incongruous sight: an old railroad coach, emblazoned with "W. A. Bechtel Co." and, in smaller letters, the name "WaaTeeKaa." Today, it houses a modest museum devoted to a century of engineering and construction accomplishments. It is also a symbol of Bechtel's enduring values. For the original "WaaTeeKaa" was the home of Stephen D. Bechtel Sr. and his family on railroad construction job sites on America's West Coast during the 1920s. Getting as close as possible to the work being done has always been a Bechtel hallmark. Stationed right outside Bechtel's headquarters, this rail car serves as a constant reminder of the ongoing commitment to doing whatever it takes to satisfy Bechtel's customers around the world.

The rail car also symbolizes Bechtel's history as a family of builders. This includes the Bechtel family, of course—four generations and 100 years of continuous family leadership of the enterprise. In a broader sense, "family of builders" refers to the hundreds of thousands of men and women who have worked in the Bechtel organization through the decades, whose collective effort has wrought the company's legacy. Their integrity, commitment, and innovation have won the trust and respect of thousands of customers worldwide, and their energy and dedication are responsible for Bechtel's success.

Today, with the values symbolized by "WaaTeeKaa" as strong as ever, Bechtel and its people are well prepared for building a new century.

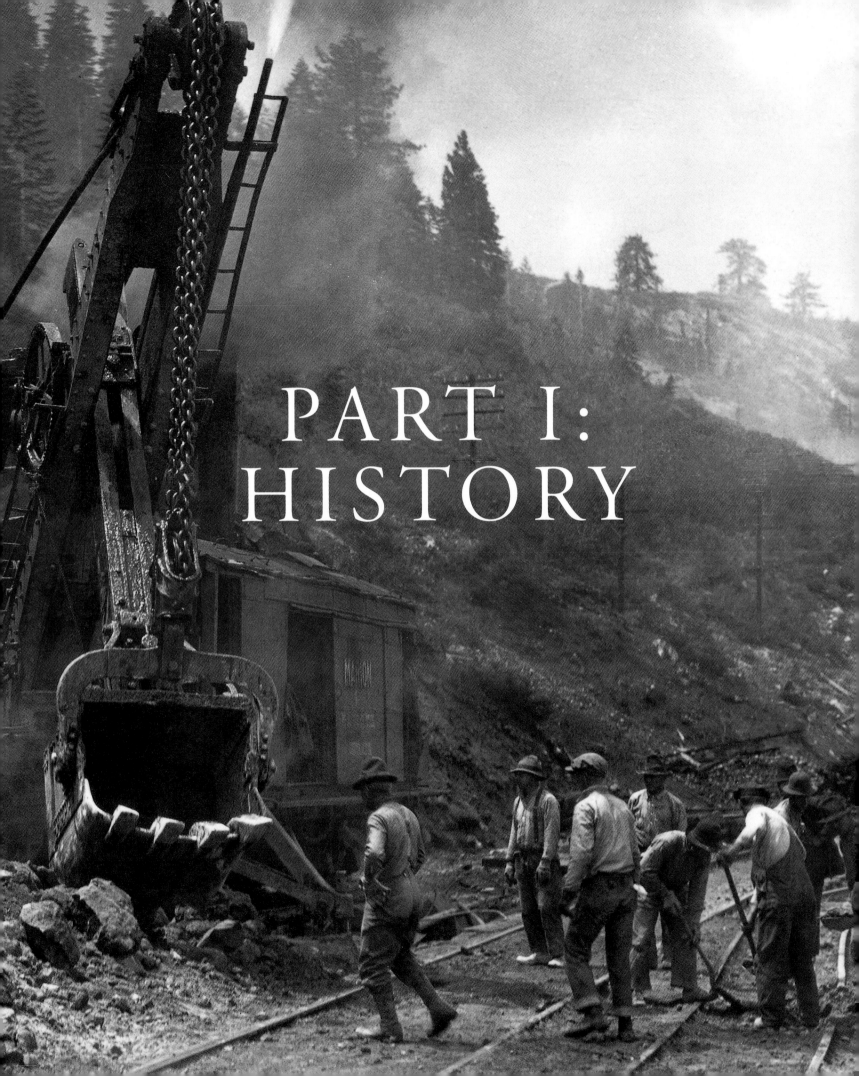

PART I: HISTORY

CHAPTER ONE

THE FORMATIVE YEARS

Building Bechtel

1898–1940

Like many great American entrepreneurs, Warren A. Bechtel would take a few frightening steps toward failure before he found the path to success. In 1898, a nearly bankrupt 25-year-old Bechtel and his pregnant wife, Clara, left Peabody, Kansas, and their cattle farm, which had fallen on hard times because of collapsing beef and corn prices, and headed 100 miles south to the dusty expanses of Oklahoma Territory in search of construction work and new opportunity. Thus began an epic journey that would span a century of building, four generations of his family, and eventually much of the globe.

BEGINNINGS: RAILROAD WORK

Although San Francisco, Los Angeles, Seattle, Portland, and Oakland—the great cities of America's Pacific Coast—were well established, vast stretches of the West remained raw and undeveloped. And while the transcontinental railroad connected the young nation, much of the West still lacked the population that could tap the natural resources of its undeveloped expanses and the network of railroads that could move crops and livestock from farms and ranches to the rapidly growing cities. By the 1890s, Southern Pacific Railroad had laid 8,000 miles of track from Portland, Oregon, south to Guaymas, Mexico, and from San Francisco east to Ogden, Utah, and south to New Orleans. Now, a host of smaller carriers were adding shorter branchlines to reach more remote regions, including the open plains of Oklahoma Territory.

1898

Warren A. Bechtel takes a job working on the railroad in Oklahoma Territory.

OPPOSITE: *In June 1927, W. A. and Clara Bechtel met the company's Caterpillar 60 en route from Bowman Lake to Emigrant Gap, in California.*

W. A. proudly emblazoned his first steam shovel with "W. A. Bechtel Co." It was used for grading at the Western Pacific site in Oroville, California, in 1909.

W. A. Bechtel became part of the migration to Oklahoma, if only temporarily. He happened to be as skilled as anyone at operating the mule-drawn sled used in railroad grading work, and labored through the spring of 1898 and the following summer in the heat, dust, and tumult of building a frontier railroad. As the new century dawned, W. A. packed up his family—his wife, Clara; their two children, Warren Jr., born December 23, 1898, and Stephen, born September 24, 1900; and his younger brother, Arthur—and headed west to Oregon. There, he became a gang foreman for Southern Pacific. He didn't stay long. When W. A. was offered a job in Wadsworth, Nevada, determining cross sections for the Southern Pacific Engineering Department, he and his family moved to nearby Reno. They didn't have a lot of baggage. "I arrived," W. A. recalled, "with a wife, two babies, a slide trombone, and a 10-dollar bill."

In 1902, W. A. was out of work and nearly out of money, but a chance encounter changed all that. He happened upon A. J. Barkley, a manager for

Southern Pacific. Barkley, impressed by the young man's manner and bearing, hired Bechtel as an estimator in Southern Pacific's Wadsworth office at $55 a month. Bechtel worked hard, learned quickly, and did well. He moved with Southern Pacific to Ogden, Utah, later returning to Nevada to serve as a gravel pit superintendent for a quarry in Lovelock. "He was learning all the time," said Barkley. "He seemed to me a natural engineer."

Barkley was not the only one to notice W. A. Bechtel's ability. Silas Palmer, an inspector for an Oakland contracting firm, recognized it immediately when he met Bechtel at the Southern Pacific gravel pit in Lovelock in 1903. "His mastery of detail was apparent," said Palmer. From the moment he took over the gravel pit, Bechtel had insisted on learning how to operate Southern Pacific's latest piece of technology: a huge, huffing, clanking Marion steam shovel. Having mastered the machinery, he understood almost instinctively how to achieve the most profitable mix of men, machines, and animals. On the spot, Palmer offered Bechtel a job working for his employer, E. B. and A. L. Stone.

By the following spring, 1904, W. A. and his family were settled comfortably in Oakland, and on July 4 of that year Karl Kenneth, his third son, was born. Dad (as the young father was now universally known by family, workers, and colleagues alike) was superintendent for the Stone company in charge of building the Richmond Belt Railroad and an extension of the Santa Fe Railroad line into Oakland. Challenging though they were, his new duties were a far cry from the rough work of punching railroad lines through the wilderness. At Stone, W. A. continued his education as a construction manager, learning new techniques to add to an already extensive knowledge base. The opportunity to see how good he would be as an independent contractor came in 1906, when he and George S. Colley Sr. teamed up to land a subcontract to make a cut through limestone on a nearby Western Pacific Railroad main line near Sunol. Because the limestone broke in irregular pieces, causing a heavy overbreak, the job was a dicey one and would have been a loser had Bechtel not demonstrated yet another skill: the ability to negotiate. He persuaded Western Pacific to take the excess limestone as fill and thereby turned a probable loss into a small profit. Flush with success, he moved quickly. He had rented a Model 20 Marion steam shovel for the Sunol job. By 1909, he owned it outright. He wasted no time in painting the name "W. A. Bechtel Co." across the side of the cab in large letters.

W. A. Bechtel lands his first construction subcontract: making a cut for Western Pacific Railroad in California. During this project, W. A. rents, and then purchases, his first steam shovel, a Model 20 Marion.

It would be another 16 years before Bechtel incorporated the business, but the purchase of the Marion was testimony to his determination to build a lasting organization that employed the latest technology. In the coming years, W. A. Bechtel Co. would constantly seek innovative ways to put technology to work and invest heavily in new machinery. Bechtel would lead the way in replacing horse-drawn freight teams, for example, with heavy, gasoline-powered trucks, typically Packards or Alcos with chain drives.

FIRST CONTRACTS

In 1909, Bechtel landed his first prime contract—grading the site of Western Pacific's Oroville, California, station on the Oakland–Salt Lake City line. Working capital was hard to come by, and the infant operation had to hustle for funds. W. A.'s finances were so thin that a friend had to guarantee his account with a wholesale grocer so he could feed his workers. He finished the job and managed to earn a small profit.

A team of mules worked at an early Bechtel job. Horses and mules were a primary source of power on construction sites in the early years of the 20th century.

While Bechtel was working at Oroville, Southern Pacific began a construction program to capitalize on the development of new traffic in the timber-rich Pacific Northwest. One of its largest projects paid $39 million to ease grades and smooth curves on the California-Oregon line. In 1910, Bechtel won a subcontract for two sections 15 miles below Natron, Oregon. W. A. personally supervised the work, using teams of horses and his cherished steam shovel. In years to come, long after he had plenty of other things to do, he would still delight in taking his turn on a scraper or running the massive Marion.

The success of the Natron project, hewed through Oregon's rugged terrain, left Bechtel with what he confided to one friend was more money than he had ever expected to have in a lifetime, and attracted the attention of the Wattis brothers, owners of the powerful Utah Construction Co. in Salt Lake City. For some time, W. A. had been pressing them for subcontracts and occasionally bidding against them on jobs. "We might as well ask him in," W. H. Wattis finally grumped to his brother, "as to have him nipping at our feet." They saw an opportunity when Ray Wattis, a son of one of the brothers, decided he was ready to take a subcontract with his father's company. The Wattis brothers gave Ray a chance, but only on condition of a partnership with the more experienced W. A. Bechtel Co.

The collaboration proved to be transforming for W. A. He came to prefer working in partnerships, which allowed him to spread risk and avail himself of specialized resources and experience. He was also a good judge of character. In hooking up with the Wattis brothers, he sowed the seeds of a relationship that would blossom in numerous subcontracts over the years. That relationship would come to full flower as a component of the Six Companies consortium that built Hoover Dam starting in 1931.

FAMILY INVOLVEMENT

W. A. Bechtel Co. was a family business from the beginning. W. A. was determined to build a company that would allow him to pass along to his children not just financial security and physical assets, but a sense of responsibility and

Supervisors and workers examined progress on the Natron cutoff in 1923, an extension of the work done 13 years earlier.

W. A. Bechtel and sons Steve, Ken, and Warren Jr. paused during construction of Southern Pacific's Natron cutoff in Oregon, 1924.

obligation to company employees and associates, and to the enterprise. He took immense pleasure in the increased interest and involvement of Bechtel family members. His younger brother, Art, was a mainstay on numerous railroad jobs; Warren Jr., Steve, and Ken all shared their father's enthusiasm for building.

There was never really any question about what the Bechtel boys would do when they grew up. Dusty equipment yards filled with massive trucks and steam shovels served as their childhood playgrounds. By the time they were in their late teens, the sons were well positioned to carry the business into the next generation. Each of them enrolled at the University of California at Berkeley and did well academically. But all left before graduation because their father needed them. Their country needed them, too. Warren Jr. and Steve shipped out to France in 1917 to serve in the U.S. Army, Warren Jr. with the 18th Engineers, Steve with the 20th Engineers.

When the boys came home in 1918, W. A. was counting on the rapidly expanding influence of a second generation of Bechtels to ensure continuity of family control. He would not be disappointed. Warren Jr., Steve, and Ken spent their time overseeing Bechtel projects and subcontracting small jobs on their own.

In 1923, young Warren took a subcontract with Utah Construction to extend his father's earlier Natron work for Southern Pacific. When the horse teams bogged down in the deep Oregon mud that winter, Warren remembered the new 60-horsepower tracklaying tractors he had seen on large commercial farms. These machines could get traction on soft ground where horses and mules were helpless. Warren bought some of these tractors, hitched trailers behind them, and moved the heavy equipment and materials. It was perhaps a first in the construction industry and would become standard practice with contractors throughout the world.

Around this time, in 1923, Steve took a job as project chief constructing Southern Pacific's main line extension into Phoenix. He had married Laura Adeline Peart earlier in the year, and when he went to Phoenix, Laura went with him. It was the first of many moves for the couple, and it established a firm policy of, quite simply, "no Laura, no Steve." If Laura couldn't travel with him, Steve wouldn't go. Their personal commitment soon became a Bechtel policy of encouraging and paying for many wives to travel on business with their husbands.

Ken's interests tended toward administration and financial management, although one early job would reveal a deeper passion. His supervision of bridge construction for the General's Highway in Sequoia National Park was the start of a lifelong commitment to natural history and the environment, reflected in the austere beauty of the masonry bridges fashioned from the region's native stone. The job lost money, largely as a result of environmental guidelines that Ken carried out to the letter. The U.S. Bureau of Public Roads required that trees be protected in blasting areas, and Ken took considerable ribbing for the metal pants he put on every tree trunk.

W. A. formally acknowledged the contribution of his family in 1925 by incorporating the company, with his sons as officers, along with brother Art. At its birth, W. A. Bechtel Co. was one of the largest and most respected construction firms in the West.

With his sons and partners overseeing much of the current business, W. A. could finally afford to devote more of his time to developing new markets. He had

Laura Bechtel brought her children to an early Bechtel construction site.

1919

Klamath Highway in northern California becomes Bechtel's first major construction project other than railroads. The Klamath job is also a California landmark: It is the first contract in the state issued by the U.S. Bureau of Public Roads.

Ken Bechtel supervised the construction of a native stone bridge in Sequoia National Park, circa 1925.

long had his eye on the power industry and was eager to find a showcase job that would demonstrate Bechtel's abilities. He found it in 1921, in the mud and waste disposal challenge of the Caribou Water Tunnel in Plumas County, California, located high up in the Sierra Nevada range near Reno, Nevada. Bechtel built a two-mile-long, 12-foot-diameter tunnel for Great Western Power, which later became part of Pacific Gas and Electric (PG&E)—the company's first job for a large power utility. Here, Steve showed his prowess, too. Working on the tunnel during a summer break from Berkeley, he devised a way to remove the waste from the excavation more efficiently and cut the project's cost significantly. He returned to school with a handsome bonus. The job provided the company not only an entrée into the power industry, but an introduction to PG&E, which would, over the years, become one of Bechtel's leading customers.

The Caribou Water Tunnel in northern California is W. A. Bechtel Co.'s first job for a large power utility. The tunnel is part of the Caribou Power Plant, which will generate 75 megawatts of electricity. A key player on the project is W. A.'s college-age son, Steve.

At the 19th century's end, Bechtel had hitched an entrepreneurial ride on the railroads as they stitched together the American frontier. As Europe edged toward the brink of war in 1914, the world in which Warren operated was rapidly changing. Railroads faced growing competition from automobiles, buses, and trucks. American transportation was diversifying, and in the decade to come Bechtel, too, branched out into highway construction and, eventually, into oil and gas production and transportation.

By 1919, Henry Ford's Tin Lizzie had become the universal car. Four million Model Ts were on the road, or what passed for roads in those days—many miles of dusty rural highways, most of them built to accommodate the horse and buggy. With the passage of the Federal Highway Act of 1921, the federal government got into the road-building business. The dominance of the railroads was ending, and the nation's lifestyle was changing. Over the next eight years, the number of miles of surfaced roads nearly doubled, providing a rich stream of projects for construction companies.

In 1919, two years before the Federal Highway Act, Bechtel won his first road-building contract: a stretch along the Klamath River in northern California. It was the first California contract issued by the U.S. Bureau of Public Roads and was followed a year later by another Bureau contract to thread a road through the San Gabriel Canyon in Los Angeles County. Art, Warren Jr., and Steve had a subcontract to do steam shovel excavation for 15 cents a cubic yard. Art ran the shovel, and Steve and Warren Jr. took turns trimming steep slopes ahead of the shovel and leading a balky mule as it pulled a four-yard western dump car. Bechtel took two more road contracts, to improve access to Sequoia and Yosemite national parks in the Sierra Nevada.

Though railroad work continued to be important for Bechtel, highway construction was becoming a substantial source of business. W. A. found the ambitious partner he needed for this booming new industry in Henry Kaiser. They first met on a road job near Redding in 1921 and later joined in the operation of a rock plant in Oroville, the forerunner of many collaborations that continued through the end of World War II.

The beautiful, arched Coos Bay Highway Bridge, a joint venture with

The Coos Bay Highway Bridge, part of the Pacific Coast Highway on the Oregon coast, is shown here under construction circa 1934.

Warren Brothers Construction Co., was Bechtel's most ambitious highway-bridge project of this period. Completed in 1934, it was the longest of five new bridges connecting sections of the Pacific Coast Highway.

PIPELINE CONSTRUCTION

In the late 1920s, in another important departure from railroad work, Bechtel entered a new field that would become a staple of the business for years to come: pipeline construction. Bechtel's first pipeline, the Tres Piños–Milpitas, which carried natural gas eight miles from Tres Piños to Milpitas, California, was built in partnership with Silas Palmer for PG&E in 1929.

Perhaps more than anyone else, Steve saw the potential in pipelines, and as the business grew, he fought to convince Continental Gas to use a Bechtel-Kaiser partnership on an upcoming project. The youthful western upstart shocked staid old Continental by bidding for the entire 400-mile line from Amarillo, Texas, to a point on the Missouri River south of Omaha. As soon as they recovered their composure, the men from Continental agreed to give Bechtel a 140-mile stretch.

Initially, W. A. and Steve had a tough time convincing their old partner, Henry Kaiser, of the promise of pipelines. But soon a Bechtel-Kaiser partnership was laying thousands of miles of pipe for Standard Oil of California, Continental Gas, and PG&E. Bechtel engineers developed a number of revolutionary techniques, including using sideboom-equipped tractors to lay pipe and welding the pipelines electrically in the field to reduce construction time.

SIX COMPANIES AND HOOVER DAM

But bigger things were in store. As the West developed, its thirst for water increased beyond nature's supply. The mountains collected ample snow most years, but the runoff rushed downriver to the sea in the spring thaw, often with devastating speed and volume. The solution was to dam rivers and create vast reservoirs behind them. Bechtel completed its first dam in 1926, for the Nevada Irrigation District. Bowman Dam was, at the time, the second-largest rock-filled dam in the

1925

W. A. Bechtel Co. is officially incorporated, with W. A., Steve, Warren Jr., Ken, and Art Bechtel listed as principals. W. A.'s growing operation specializes in railroad projects, including grading railroad beds, enlarging tunnels, and constructing train sheds and snowsheds.

Bechtel workers used new techniques for coating pipelines directly in the field, which substantially reduced construction time.

Bechtel's first foray into
the dam-building business,
Bowman Dam, is completed.
The construction site is so
remote that the company is
forced to construct a camp,
complete with a hospital, a
hundred head of cattle, a
slaughterhouse, and storage
facilities, to sustain the crew
for the winter.

Bechtel's first pipeline,
the eight-mile-long Tres
Piños–Milpitas in California,
begins operation. This
entry into the pipeline
business is championed by
Steve Bechtel; over the next
six years, the company lays
more than 1,000 miles
of pipe.

world, in addition to being Bechtel's most remote job site, in the center of California's rugged landscape. "We were really cut off from civilization," Warren Jr. recalled later. "We had to get everything in there to carry us from Christmas until the end of April." It was a forerunner of many projects to come in remote outposts. The company completed the dam, even though the district ran out of money to pay for the whole job.

Although they were building projects on a huge scale, western firms were still thought of as regional operators by the powerful, old-line eastern construction companies and their customers. The westerners longed to change that perception and had begun to lay the groundwork for an audacious undertaking. A consortium of western contractors was quietly assembling a team to undertake the most challenging project ever built in the United States—damming the mighty Colorado River, something engineers had dreamed about for decades.

Oddly, the onset of bad times would provide the impetus to go forward with the Colorado River project. W. A. Bechtel Co. was barely four years old on October 24, 1929, when the stock market crashed, plunging the nation into the Great Depression. In 1930, as the economy sank, Congress approved a public works project at Black Canyon, Nevada, the future site of Hoover Dam. The project would tame the flood-prone river, protecting cities and farms; it would generate cheap electricity to supply power to homes and industry; and it would provide work, thousands of desperately needed jobs in occupations ranging from ditchdigging to geology.

The Hoover Dam project was too big for any one company. So W. A. Bechtel helped form a consortium calling itself Six Companies, Inc. W. A. knew the heads of the consortium companies as friends and business associates, having been in partnerships with most of them. There was tall, lean Harry Morrison, head of Morrison-Knudsen of Boise, Idaho, and the man most directly responsible for bringing the group together; and the white-haired Wattis brothers of Utah Construction Co., the region's foremost railroad builders. They were joined by the wry Felix Kahn of MacDonald & Kahn, a premier builder of office buildings, industrial plants, and hotels, including the Mark Hopkins in San Francisco. Phil Hart ran Pacific Bridge Co., one of the oldest construction firms on the West Coast, and was justly famous for his underwater work—a critical component in dam construction. Charlie Shea, the pugnacious, acid-tongued boss of J. F. Shea

Co., was the best tunnel and sewer man west of the Rockies. And finally there was the legendary Henry Kaiser, whom W. A. had long valued for his enthusiasm and vision. W. A. Bechtel served as the second president of Six Companies; his son Steve was a member of the executive committee; and sons Warren and Ken served on the board.

Assembling the consortium was easy; making a bid for the largest labor contract ever awarded by the United States government was the estimating equivalent of working without a net. Damming the Colorado River at Black Canyon would be a fearsome technical and organizational challenge. "The site of the dam," noted *The New York Times*, "is in one of the wildest and most inaccessible parts of the United States. It is a furnace in the summer and often bitterly cold in winter."

Six Companies and U.S. Bureau of Reclamation officials were proud participants in the construction of Hoover Dam. From left: H. J. Lawler, W. R. Young, Charles A. Shea, E. O. Wattis, Elwood Mead, Frank Crowe, R. F. Walter, and W. A. Bechtel.

A group including Bechtel
begins constructing Hoover
Dam. Completed in 1936,
two years ahead of schedule,
the project includes four
of the largest rock tunnels
ever built, the dam itself,
and two giant spillways,
each large enough to hold
a battleship.

Construction work on the dam, the biggest civil engineering project ever undertaken, would have to be finished in seven years. During that time, 3.7 million cubic yards of rock had to be excavated, 4.4 million cubic yards of concrete poured, and 45 million pounds of pipe and structural steel erected. The result would be a massive arch-gravity dam nearly a quarter of a mile long and more than 70 stories tall. Despite the consortium's meticulous calculations and decades of combined experience, a fixed-bid job of this magnitude was a very risky proposition.

Though dozens of firms had considered the mammoth job, only five submitted formal bids. These were announced at a raucous gathering in Denver on March 3, 1931. The first two were thrown out: one for "$80,000 less than the lowest bid you get" and a second for $200 million or "cost plus ten percent." Lacking the required bond, the pretenders were quickly dismissed. Arundel Corp., one of the old eastern giants, put in a $53.9 million bid. The fourth bid, from Woods Brothers of Lincoln, Nebraska, was for $58.6 million. Finally, Six Companies' bid of $48,890,955 was announced. W. A. Bechtel and his partners had bid $5 million less than the next highest bidder and just $24,000 more than the cost calculated by Bureau of Reclamation engineers. Six Companies of San Francisco would build Hoover Dam.

Work began at a furious pace, outstripping anything anyone could have expected. As a result, as 1931 drew to a close, W. A. Bechtel announced, "We are at least eight months ahead of schedule." Four months later, little more than a year after construction had begun, a beaming W. A. announced that work on Hoover Dam was a full year ahead of schedule. Six Companies crews had accomplished two years' work in the space of one.

W. A. didn't live to see the historic work completed. In 1933, with construction activity on the dam at its peak, W. A. decided it was "a good time to see what the rest of the world is doing." Confidently leaving the nation's most ambitious civil engineering project in the capable hands of his partners and his sons, he and Clara accepted an invitation from the Soviet government to visit the just-completed Dnieprostroy Dam. It was an opportunity to compare notes on men and methods on the world stage, and W. A. was determined to take it. But on August 28, 1933, 15 days before his 61st birthday, he died suddenly in Moscow. It was a great shock to his family and friends, but they took solace in the knowledge that he went as he would have wanted to go, active to the last, a construction

OPPOSITE: *Hoover Dam was
under construction in 1934.*

Workers at Hoover examined a cargo of dynamite for blasting diversion tunnels. At its peak, the project employed 5,000 workers.

man on a construction mission. "There was nothing to do but close ranks," said Steve Bechtel. "And we did." Steve was elected to succeed his father as president of W. A. Bechtel Co.

On balance, though, fate would prove kind to Six Companies. First the Depression helped make Hoover Dam possible; then it cut costs. Economic conditions worsened in the early 1930s, causing severe price deflation of everything from milk to cement, and the cost of materials dropped further as work progressed. Nature provided a second, backhanded assist. A devastating drought that choked the West in the great Dust Bowl reduced the flow of water in the Colorado River. The water level, lower than it had been in a decade, made the crucial work of rerouting the river to the cofferdams and diversion tunnels far less difficult than had been expected.

On September 30, 1935, President Franklin D. Roosevelt officially dedicated the dam, saying, "I came, I saw, and I was conquered, as everyone will be who sees for the first time this great feat of mankind." Five months later, a lone newsreel camera followed Superintendent Frank Crowe as he walked to the crest of the dam to announce the end of construction—two years, one month, and 28 days ahead of schedule.

Hoover Dam represented a pivotal event in the history of Bechtel. There have been bigger projects in recent years, and there will be still bigger jobs in years to come. But never again will Bechtel be involved with a project that so profoundly shapes its sense of itself. As Stephen D. Bechtel Jr. said in 1982: "Hoover was a make-or-break proposition for my grandfather. Hoover Dam became the birthplace of many of the great traditions of the present Bechtel organization. There was skilled management, individual initiative, innovative thinking, risk-taking, and, above all else, a kind of driving determination to make the project successful." And as Steve Sr. would note years later, "Hoover Dam put us in a prime position as being big-time, real thinkers. We bet our shirts on it."

BEYOND HOOVER DAM

In 1933, with unemployment reaching 25 percent, President Roosevelt was committed to restarting the economy. Through the Works Progress Administration and other agencies, he launched billions of dollars' worth of large-scale infrastructure projects, including the building of roads, bridges, and dams.

Hoover Dam workers hung suspended in a cable car hooked to a cable over the vertiginous 2,000-foot canyon.

Bechtel, as part of the Six Companies consortium, was ideally positioned to take advantage of the public works opportunities. The consortium's monumental work on Hoover Dam demonstrated a construction prowess that few could match. "No group of Americans in history have spread themselves out on the scale of these westerners," asserted the editors of *Fortune* magazine in a 1958 series called "The Earth Movers." "They are the instruments with which the West is progressing confidently and irresistibly toward its goal of empire."

Between 1934 and 1938, the partnership built Parker Dam across the Colorado River; operating as Columbia Construction Co., they corralled the waters of the Columbia River behind the concrete arches of Bonneville Dam and built Ruby Dam and Grays Harbor jetties in Washington State; and as Six Companies of California, they constructed a section of the Oakland–Contra Costa highway. Utah-Bechtel-Morrison-Kaiser Co. lined the Moffatt Water Tunnel and built Taylor Park Dam in Colorado.

1933

Steve Bechtel becomes vice
president of Bridge Builders,
Inc., a joint venture of
construction companies that
takes on the job of sinking
foundations for the eastern
cantilever span of the
San Francisco–Oakland
Bay Bridge.

SAN FRANCISCO–OAKLAND BAY BRIDGE

Reaching across San Francisco Bay in two sweeping spans that connect on Yerba
Buena Island, the Bay Bridge, started in 1933, would link San Francisco with the
city of Oakland. Not only would it make commuting between the two cities
feasible, but it would also connect the West's financial center, San Francisco, with
industrial Oakland.

These were times when attempting the impossible became standard procedure, and the Bay Bridge is a perfect example—the plans called for bridge piers to
be set deeper than anyone had ever attempted. And seldom had anyone confronted
currents as unpredictable or winds as blustery as those on San Francisco Bay. The

Bechtels formed Bridge Builders, Inc., allying themselves with Six Companies partner Henry Kaiser, one of W. A.'s most trusted collaborators. They added several old-line eastern firms, including some of the most experienced deep-caisson builders Steve could find.

Bridge Builders won the contract for elements of the bridge's eastern portion, and Transbay Construction, which included other Six Companies partners, won the contract for the western portion. A lively rivalry developed between the old Hoover Dam partners as they labored on opposite sides of Yerba Buena Island. Transbay at one point claimed the record for the world's deepest pier, set at minus 246 feet. But Bridge Builders countered with one of its own. When it was over, the

The San Francisco–Oakland Bay Bridge, under construction in the mid-1930s, was the longest bridge in the world at that time, extending 8.2 miles shore to shore.

Industrial Engineering Co.,
a Bechtel subsidiary, is
established to manufacture
and apply a Somastic pipe
coating, which reduces
corrosion and greatly
increases a pipeline's
lifespan.

*Steve Bechtel's interest in
pipelines spurred the company
to become the leader in
the construction of large
pipelines during the late
1920s and early 1930s.*

Bechtel combine had the world record, setting its deepest pier at minus 247 feet. The bridge was completed in 1936.

PIPELINE BUILDING AND PETROLEUM REFINING

For several years, Steve's interest in pipelines had been growing, along with his interest in a closely associated and increasingly important field, petroleum and its by-products. Neither W. A. nor Henry Kaiser shared his interest, and Steve spent hours trying to convince them of the promise of pipelines. Finally, they came around, just in time to catch a pipelining boom.

Steve's prescience about pipelining was no fluke, and over the years his instincts would be proven right again and again. Bechtel combined technological leadership and innovative organization to drive the company into the pipeline industry's top tier. The company was the first to use a sideboom tractor to lay pipe, allowing it to turn the traditional piece-by-piece method into a more or less continuous process that cut costs substantially. In 1936, Steve enhanced Bechtel's edge by setting up a subsidiary called Industrial Engineering Co. to manufacture and apply a revolutionary new pipe coating, called Somastic coating. The process reduced corrosion and extended the normal 10-year life of a pipeline to 50 years.

In 1936, Steve and Ken reorganized the company. W. A. Bechtel Co. would continue to handle the traditional earthmoving work, and a new firm, S. D. Bechtel Co., was set up to pursue other lines of business. For some time, Steve had been making inquiries among western oil refiners and discovered that, as he had suspected, all of them depended on construction firms based east of the Rockies. There was clearly a market for a broad-based firm in the West that could provide engineering, management, and construction services for petroleum refining and processing.

To tap this emerging market, S. D. Bechtel would have to hire the best talent in the field. Steve had a pretty good idea where he could find it. As director of purchasing for Hoover Dam, suppliers came to know Steve as "the man to see." One of those who had been sent to see him was John McCone. A brilliant, hard-driving businessman, McCone had been a salesman with Consolidated Steel of Los Angeles and had successfully bid for some of the steelworks at Hoover. Steve had known him during their student days at Berkeley. "Steve and I shared a sense

of imminent change," McCone would recall, "of great projects about to break at last upon the West. We were sure we could have a place in them." The last piece of the new company came into place in May 1937, when Bechtel and McCone brought in a 43-year-old refinery designer from Chicago named Ralph Parsons.

S. D. Bechtel Co. re-formed that year into Bechtel-McCone-Parsons Corp. (BMP) with a capitalization of $75,000. And that year, BMP built its first refinery, a hydrogenation plant for petroleum refining, contracted by Standard Oil of California in Richmond. This marked Bechtel's move into engineering. Steve was so sure that the petroleum industry needed a complete refinery engineering and construction firm in the West that he offered to do the job for free. Construction began in the summer of 1937 and was completed the following January. By late 1941, BMP had eleven refineries completed or under way.

A section of pipe for a natural gas line was lowered by Caterpillar tractors with Bechtel & Palmer homemade sideboom winches in 1930.

At the same time, Steve was pushing hard to involve Bechtel in foreign work. He succeeded in 1940, with the 75-mile, 16-inch Mene Grande pipeline for Standard Oil of Venezuela. The project included 40 miles of roads, 132 miles of telephone lines, and harbor and wharf installations for the Venezuelan government at Ciudad Bolívar.

With Steve's thoughts focused on petroleum, it would be only a matter of time before he began looking at the Middle East. The area was awash in oil, but bringing it to market would require ocean transport. Quietly, Steve had ordered a study of shipbuilding that confirmed his suspicion that the industry was on the verge of taking off. He carried a summary of the study in his suit pocket and would pull it out whenever he lunched with one of his former partners from Six Companies. But, with the exception of Henry Kaiser and Charlie Shea, he could not get them interested. They were reluctant to sign on without a guaranteed buyer in sight. They would not have a long wait to see one appear.

1937

Bechtel-McCone-Parsons Corp. is formed, specializing in petroleum refinery engineering and construction. Steve Bechtel, who antici-pates rapid growth in the oil business, heads the new corporation.

1937

Bechtel-McCone-Parsons wins a contract to engineer and build a hydrogenation plant in Richmond, California. The plant is BMP's first refinery, as well as one of the first petroleum refineries in the West.

1940

The 75-mile Mene Grande pipeline in Venezuela is Bechtel's first project overseas. The pipeline transports oil from the interior to tankers on the coast.

OPPOSITE: *A 140-ton butane column was designed for Standard Oil of California's El Segundo refinery.*

WARREN A. BECHTEL

"W. A.". . ."BECK". . .
"Warren." W. A. Bechtel responded to a number of nick-
names over the years. But the
one that really stuck was
"Dad." His wife, Clara,
bestowed the name on him
when their first child, Warren
Jr., was born in 1898. Soon fellow workers at construction
sites began calling him Dad,
too. The name was appropriate. Dad was a natural leader,
a reliable father figure who was known even in construction camps as a good provider.

Warren A. Bechtel was born on September 12, 1872,
the first of Elizabeth and John Moyer Bechtel's seven children. He was raised on the edge of the frontier in Freeport,
Illinois, on a farm 25 miles east of the Mississippi. When he
was 12 the family moved to Peabody, Kansas, where they
ran another farm as well as a grocery store.

Warren was a restless, energetic teenager who found
time for school, farm and store chores, and the slide trombone. After graduation, he had a brief fling as a traveling
musician, but realized quickly that he couldn't make a living with his trombone.

In the spring of 1898, Warren was 25, Clara was pregnant, and his cattle ranch was nearly bankrupt. He heard
that the railroads in Oklahoma would pay a man with his
own team $2.75 a day to grade track beds. The man who
would break new ground in construction techniques was for

*The Bechtel family circa 1915. From left, standing: Clara,
Warren Jr., Steve; seated: Ken, Alice, and Dad.*

the moment content to break
the prairie to put bread on the
table. Warren and his wife left
Kansas for the open, undeveloped plains of what was called
Oklahoma Territory with only
a pair of mules, the last assets
from the ranch.

In Oklahoma, he found
his life's work. The stint on the
railroad led to every kind of
construction project, from
snowsheds to pipelines to
Hoover Dam, and to the global company that bears his
name. Yet the essence of Warren Bechtel was that he loved
the physical labor of construction, and long after he had to,
he would take a turn with a scraper. "He used to climb up
on a steam shovel and load a couple cars, enjoying every
minute of it," recalled a coworker. "W. A. came right up
from the grass." A self-taught student of engineering and
business, he never stopped learning. "I've always thought
good engineers are born, not made," said A. J. Barkley, one
of his early supervisors. "They must have a knack for it.
Beck is what I've always thought an engineer should be—a
man who understands what is to be done, knows how to do
it, and finishes the job economically."

From 1898 to 1906, Dad chased the railroads west,
working on the network of branchlines threading through
the region. At each job—gang foreman, estimator, gravel
pit manager, superintendent—Dad would add to his growing collection of skills.

As the wife of an itinerant construction manager, Clara Bechtel developed her own skills. In the early days, Dad's jobs sometimes took them to the ends of civilization. Although raised in a well-to-do family, she delighted in discovering that she could fashion a home just about anywhere, even in a railroad camp tent or a converted boxcar.

The family was quite comfortably ensconced in Oakland, California, in 1906 when Dad made his first venture toward independence. With a partner, George S. Colley Sr.—a coworker and friend—and a rented steam shovel, he took a subcontract on a Western Pacific Railroad project. The job made only a small profit, but Dad made two important resolutions. The first was that he would always use the most modern equipment and techniques available, a decision of critical importance. The second was that, whenever possible, he would collaborate with others; he would find and work with many partners over the years.

W. A. ("Dad"), far right, visiting the crew on the site of an early Bechtel project.

By 1912, Bechtel had assembled a team that included his brother, Art, and George Colley Sr. Together, they completed three small subcontracts and then Bechtel's first big job, a 106-mile link for the Northwestern Pacific Railroad through the Eel River Canyon, which was completed in 1914.

Dad wanted Bechtel to be a family company and relished the prospect of handing the company over to his sons. Warren Jr., Steve, and Ken (along with sister Alice) grew up watching their father build railroads from 1906 through 1917. All three boys attended the University of California at Berkeley, but each left before graduating to join their father, because, as Steve put it, "Dad needed us in the business."

In the 1920s, construction became a different business because of the postwar boom fed by increases in electric capacity and automobile ownership. Bechtel shifted with the times. In 1919, Dad received his first federal highway contract, and it was quickly followed by other highway jobs. In 1921, he built the Caribou Water Tunnel in the Sierra Nevada, his first work for a power utility. The Bechtel boys played an increasing role in managing these and other projects, to the point that when Dad formally incorporated W. A. Bechtel Co. in 1925, he made his sons and his brother, Art, officers in the corporation. The next year would mark another first: construction of Bowman Dam, at the time the world's second-largest rock-filled dam.

Although successful, Dad had longed to make the company a player on the national scene. In 1931, he got his chance; as part of Six Companies, Inc., a consortium of contractors, Bechtel was involved in the largest civil engineering project in history, Hoover Dam.

Early in 1933, Bechtel participated in the creation of Bridge Builders, Inc., constructors of the piers supporting the eastern span of the San Francisco–Oakland Bay Bridge. Later that year, with construction of Hoover Dam moving along at a brisk pace, Dad and Clara accepted an invitation from the Soviets to visit their country. Dad died suddenly on August 28, while in Moscow. He left behind a portfolio of awe-inspiring work. But the most enduring monument Dad Bechtel built was not a dam or a railroad; it was the ideals embodied in the family firm that would bear his name into the next century.

CHAPTER TWO

FROM DAMS TO INTERNATIONAL CONFLICT

Bechtel Goes to War

1940–45

I n the decade leading to World War II, Bechtel had weathered the Depression and in the process was able to increase its expertise in engineering, design, and construction. When war came, the company not only would make a massive contribution to the production effort but would also devise many manufacturing innovations.

CALSHIP AND MARINSHIP

War engulfed Europe in September 1939. An American industrial machine would ultimately bury the Axis powers beneath a mountain of matériel: airplanes, jeeps, artillery, tanks, ordnance, and, of course, ships—ships that Steve Bechtel wanted very much to build.

Unknown to him, in the summer of 1940, the United States was anxiously seeking ways to expand its shipbuilding capacity. Bechtel-McCone-Parsons, through a joint venture with Todd Shipyards, had already won an order from the U.S. Maritime Commission for five C-1 cargo ships. But not long afterward, the British placed an urgent order for 60 more C-1s, to replace the tonnage that German submarines had sunk. Through a chance meeting with a friend, John McCone learned of the Maritime Commission's needs and arranged a meeting with Admiral Howard L. Vickery, the U.S. Navy officer responsible for naval ship design, construction, and purchasing.

OPPOSITE: *Ships were launched with remarkable frequency from Marinship. West Coast wartime emergency shipyards run by Bechtel interests were models of innovation and productivity.*

While Calship turned out cargo vessels at a record rate, Marinship specialized in building tankers and oilers to more exacting specifications.

Vickery offered Bechtel interests the opportunity to build 30 ships, but Steve insisted on bidding for all 60, despite never having built so much as a dinghy. It would be an enormous challenge, particularly since the shipyards would have to go into production instantly and would have to mass-produce ships from a single design, something no one had ever done before.

Hoover Dam had given Bechtel an appreciation for economies of scale, and Steve was certain that volume would work in his favor. "If we had all 60," he said, "we could control the market for the materials and equipment necessary to do them rather than have three or four different yards competing for the steel and the engines and the gear. Size can work to your advantage if you think big. You just recognize it and move the decimal point over. Instead of taking 2,000 people, it would take 20,000, or 50,000."

Bechtel won the contract, from which would come two of the most productive shipyards of the war. First came Calship in 1941 at Terminal Island in Los Angeles harbor. By war's end, Calship had produced 467 vessels, among them the cargo vessels called Liberty ships, the speedier Victory ships, and tankers. In the fall of 1944, Calship was turning out 20 ships a month, operating three shifts a day, seven days a week, making it perhaps the most productive shipyard in history. The last ship would slip into the water on October 27, 1945.

Barely three months after Pearl Harbor brought the United States into the war in December 1941, Bechtel received an urgent telegram from Maritime Commission Chairman Admiral Emory S. Land, requesting a new yard—one that could start producing ships before the end of the year. Within days, Marinship, managed by Steve's 38-year-old brother, Ken, was taking shape across the San Francisco Bay in Sausalito. It, too, would become a wartime production marvel, cranking out 62 T-2 tankers, 15 Liberty ships, and 16 oilers by war's end, a production level considered outstanding given the difficult task of building tankers.

Marinship launched its first vessel in 126 days—51 days ahead of schedule, and twice as fast as other new yards. Workers had barely celebrated their impressive start in building Liberty ships when they were ordered in 1943 to shift to production of T-2 tankers. Instead of mass-produced cargo ships, Marinship would build the larger, more powerful, more technically exacting tankers. These vessels would be fast enough to run with the fleet, allowing large roving U.S. Navy task forces to leave base for months at a time and carry the fight to the enemy's home waters.

Necessity produced other firsts. For the first time, large numbers of women joined the workforce, filling jobs traditionally held by men. Marinship had the highest percentage of female workers of any yard in the nation. Women at Marinship were joined by minorities who were also getting their first chance at skilled jobs historically closed to them. Employees of all colors worked side by side, earning equal pay and equal benefits. "There was no such thing as segregation," recalled Thelma McKinney, an African-American worker. "People were so busy they didn't have time for racism."

Then, in October 1945, after nearly five grueling years of round-the-clock production, Marinship's and Calship's work was done. The Maritime Commission asked Bechtel to stay on and operate the yard as a government-owned plant, but Ken turned them down. At midnight on May 16, 1946, he turned Marinship over to the U.S. Army Corps of Engineers as an operations center for its Pacific Islands Reconstruction Program. "Nothing had outwardly changed," Ken later wrote, "but I realized that Marinship as we knew it had become history." And quite a history, at that.

Steve Bechtel and Saudi Arabia's Prince (later King) Faisal Bin Abdul Aziz toured Marinship in 1944. The Bechtel–Saudi Arabia relationship was important in the coming years, as the Middle East petroleum industry developed.

1941-45

Calship delivers 467 cargo ships; Marinship cranks out 15 Liberty ships and 78 tankers and oilers. The ships are built at breakneck speed by engineers, constructors, and workers—most of whom have no previous experience in shipbuilding.

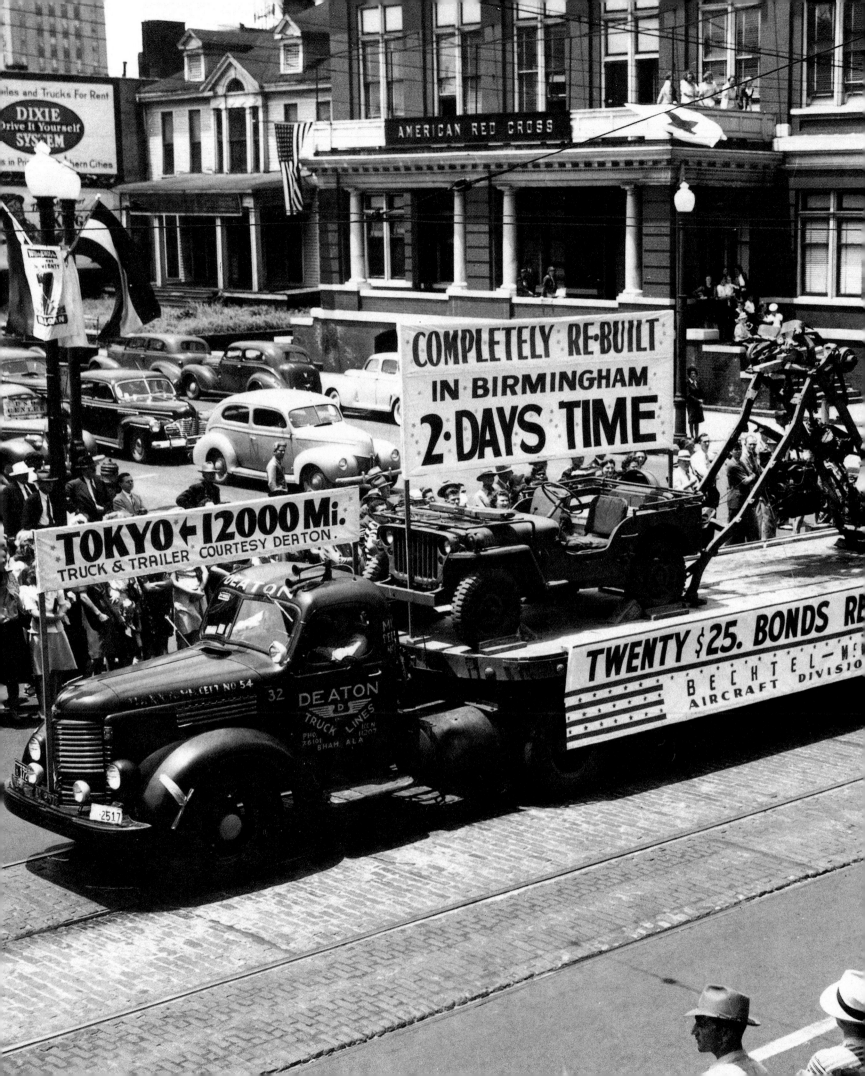

MILITARY BASES AND OTHER WAR PRODUCTION

Bechtel's war duty was not limited to ships. A Bechtel concern built and operated a sprawling plant, the Aircraft Modification Center in Birmingham, Alabama, where Perry Yates oversaw 14,000 workers who modified B-24 and B-29 bombers for specific duties. The center also hired a group of 600 men with only moderate levels of skill to restore battle-fatigued jeeps. With the knowledge gained from shipbuilding and airplane modification, managers oversaw the restoration and modernization of thousands of vehicles. With Standard Oil, the company operated Pacific Tankers for the U.S. Navy, a fleet of 90 ships that was one of the largest oil movers in the world. Bechtel also built military bases, from the Naval Air Station at Corpus Christi, Texas, to Fort Ord near Monterey, California, to Elmendorf Air Base in Alaska. It built explosives plants in Missouri and New Jersey, copper mines in Arizona and Mexico, and the Oakland Army Depot, through which munitions were shipped to forces fighting in the Pacific.

THE TRAGEDY OF CAVITE AND SANGLEY POINT

Since May of 1939, Bechtel had been one of eight contractors working on the Pacific Naval Air Bases program to link a chain of bases stretching over more than 10,000 miles, from Alameda, California, to Pearl Harbor, Midway, Wake, Guam, and the Philippines. The goal was to strengthen American air power in the Pacific.

But with the Japanese invasion of Indonesia in the early part of 1941, time was running short to get American protection in place when Bechtel signed on for the dangerous job of expanding facilities in the Philippines. Steve put his trusted lieutenant, Vice President George S. Colley Jr. (the son of one of W. A.'s first employees), in charge. "He can get people to do what they don't know they are capable of," Steve said.

Colley arrived in Manila on April 20, 1941, to learn that there were no blueprints, no site plans—just a list of jobs the U.S. Navy had to get done if it was going to build up a defense at Cavite on Manila Bay, and at nearby Sangley Point. The list included dredging and building seaplane hangars and building fuel and ammunition storage facilities, a powerhouse, and living quarters. There was plenty

Bechtel expands operations to Asia, as a team is dispatched to Manila Bay to help reinforce naval defenses. War overtakes the effort, with one of the company's top executives, George S. Colley Jr., captured and interned for the duration of the war.

Bechtel interests establish the Aircraft Modification Center in Birmingham, Alabama, where new planes are adapted to even newer requirements. This process avoids sweeping changes to the original production lines.

War bonds were sold to finance the remanufacturing of jeeps at the Aircraft Modification Center. A parade helped boost patriotic spirit.

of labor, but Colley's team would have to scrounge for materials and equipment. Turned down by Philippine officials, they went to President Manuel Luis Quezón himself for permission to use the only dredge in the islands. Getting steel was the next problem. Somehow, they managed to slip a load of steel out of Hong Kong under the noses of the Japanese. All this work didn't go unnoticed. Bechtel received some kind words from General Douglas MacArthur, who was charged with defending the Philippines. "It does my heart good to hear about the plans you and Colley are carrying out," he told the Bechtel team. "I pray to God you will finish in time—but I do not think you will."

Sadly, MacArthur was right. While the American fleet was being decimated at anchor in Pearl Harbor, Japanese dive-bombers were also striking Clark Field and Manila, destroying American air power in the Philippines on the ground. In the days that followed, bombers returned at will to complete the destruction of Cavite and Sangley. It was a devastating personal loss for Steve Bechtel. Thirty-five Bechtel men were among those who died or were taken prisoner by the Japanese.

After the fall of Cavite, Bechtel crews sought refuge in Manila. Steve called George Colley every night—to offer moral support as much as anything else. But on December 28, knowing the Japanese had tapped the line, Colley signed off by saying, "You won't hear from me again for about a month." A week later a cable arrived with the news: Colley and a small party had escaped by sea. Guiding a commandeered pleasure launch through minefields and enemy waters, Colley, his wife, Marjorie, and his crew traveled by night and hid by day, making their way across the Mindoro Strait to Balabac, where they switched to a small native sailboat called a *kumpit*. They reached Borneo in their storm-battered craft and hid out in a swamp hoping to find a more seaworthy vessel for the daunting 1,000-mile journey to Australia. George was swimming to a native village in search of a boat when he encountered a crocodile and had to scramble up a tree to safety.

A Japanese patrol boat plucked him from his perch, and he and his companions spent the next three years in prison camps. Colley recorded his experiences in *Manila, Kuching and Return*. On September 12, 1945, Colley was reunited with his wife, who had been held captive in another camp. Steve Bechtel would join them in Manila for a moving reunion. "We sat up most of the night," recalled Colley, "trying to piece together some of the happenings of the last four years." Among the stories Steve told was the epic of the Canol pipeline.

At the Aircraft Modification Center in Birmingham, Alabama, 5,000 battle-scarred jeeps were salvaged from the junk pile to be revived and modernized in only eight months.

OPPOSITE: *Workers reconstructed a damaged World War II B-29 bomber at the Aircraft Modification Center.*

Workers at the Canol pipeline in the frozen wilderness 75 miles south of the Arctic Circle did not always return to camp for their midday meal; instead, they lunched along the road.

CANOL

Canol was a top-secret project, the details of which were known only to a small circle of senior corporate officers and key defense officials. Japanese troops had already invaded the Aleutian Islands, just 1,200 miles west of mainland Alaska, and the United States feared that they would establish submarine and air bases that would cripple shipping in the north Pacific. A secure oil supply would be essential to Alaska's defense, but before the war there wasn't so much as a road connecting the area with neighboring Canada.

In the spring of 1942, the joint venture Bechtel-Price-Callahan, by direct command of Secretary of War Henry Stimson, was ordered to build a 1,430-mile pipeline spanning portions of Alaska, the Yukon, and unmapped portions of the Northwest Territories, described by one writer as a place where "the mountains are nameless and the rivers all run God knows where."

The pipeline began 75 miles below the Arctic Circle, at Norman Wells, Northwest Territories. From there, it went 580 miles southwest to a refinery at Whitehorse in the Yukon, which would supply fuel to 10 airfields in the area as well as for other military needs. Bechtel would also have to make the region habitable by building roads, airstrips, housing, and communications centers, starting from a base camp north of Edmonton, Alberta.

It was an awesome task and a logistical nightmare. Not only did the pipeline have to be built, but branchlines, pumping stations, a refinery, a tank farm, and storage facilities were needed, not to mention a way to get all the equipment and personnel in place. Until the 1920s, the only means of transportation in the region were dog teams in winter and boats in summer. Bechtel used a system of inland waterways to move food and equipment north; a chain of airfields was built so that the engineers could hop quickly up and down the line.

The working conditions were miserable to start with and then got worse. In winter, there was snow and ice and subfreezing temperatures; in summer, there was three months of continuous daylight, accompanied by mud, dust, and clouds of flies and mosquitoes. It's no wonder that 20,000 people had to be recruited to maintain the workforce of 4,000 that was needed to finish the job.

During nearly two years of construction, Steve became a subarctic commuter, staying in regular touch with field crews in the most remote outposts. Eventually, the heavy demands of travel to the Arctic, the Pacific, and the Middle East, the 18-hour days, the seven-day weeks, and the never-ending need to stay on top of dozens of complex projects took their toll. By the time peace came in the fall of 1945, Steve was physically exhausted, and Bechtel's great burst of adrenaline-fueled wartime production came to an abrupt end.

1942

As part of a Department of War–mandated project to shore up U.S. defenses in Alaska, Bechtel interests begin work on Canol, a 1,430-mile pipeline across Canada and Alaska. The pipeline transports oil from Canadian fields to a new refinery.

1943

Bechtel's first work in the Middle East: Bahrain Petroleum Co. hires BMP to enlarge a refinery to produce 150,000 barrels of oil per day.

1944

Arabian American Oil Co. (Aramco) hires Bechtel to build a small refinery at Ras Tanura. It is the first of what will be decades of projects in Saudi Arabia. Fifty years later, in 1994, Bechtel is selected to upgrade the refinery.

STEPHEN D. BECHTEL SR.

STEPHEN DAVISON BECHTEL'S success flowed less from his easygoing optimism than it did from the qualities that lay just beneath that freewheeling, affable appearance—a strong self-discipline, an iron will, and a driving determination to get things done. "There's a saying that a construction company is the lengthened shadow of one man," said John L. Simpson, Steve's longtime financial adviser. "If ever that was true, it was true in this case. Warren Bechtel was a very successful businessman. But the man who really dreamed great dreams and put them into effect was Steve. Steve Bechtel must have got out of his crib determined to do something active and important."

Raised in construction camps, he spent his teenage summers working with construction gangs. After graduating from high school, he shipped out to serve 19 months in World War I, burning up the French countryside as a motorcycle dispatch rider with the 20th Engineers. On his return, he resumed his education, studying engineering at the University of California at Berkeley. But he dropped out after his junior year to join the business.

A year later, at the age of 22, in September 1923, he married his college sweetheart, Laura Adeline Peart. They moved in across the hall from Mom and Dad Bechtel in

Stephen D. Bechtel, on motorcycle, during his World War I service in the U.S. Army 20th Engineers.

the Art Deco apartment building W. A. and partners had built near Lake Merritt in Oakland, California. On May 10, 1925, Stephen D. Bechtel Jr. was born, and not long after that Laura gave birth to a daughter, Barbara.

By the time Steve Sr. reached his late 20s, he was in effect the CEO of Bechtel-Kaiser's joint operations, and was eagerly pressing Dad and Henry Kaiser to get into the pipeline business. An early and ardent advocate of company expansion, Steve wanted W. A. Bechtel to pursue a more diverse workload, and to reach out geographically to broaden what was essentially a regional concern.

When Dad Bechtel died unexpectedly in 1933, in the middle of the Hoover Dam project, there was little doubt who would take his place. Steve Bechtel, then 32, was already on the executive committee of the Six Companies consortium building the dam, in charge of all administration, purchasing, and transportation. "I was temperamentally more suited to take on the lead," said Steve. "Warren, Kenny, and I had several talks, and Warren just didn't want to work like we wanted to work."

A moderate man, Steve went to bed early, politely excusing himself from most social functions around 10 o'clock. By 7:30 A.M., he was usually at his desk at 155 Sansome Street—that is, when he was in town. He and

Laura spent six months a year visiting Bechtel projects around the world.

Steve was a hands-on manager, yet he ran a loose, informal organization that suited his relaxed style. Until Simpson came on board in 1942, Steve received all the financial advice he needed from Bechtel's attorney, Robert Bridges, and an accountant named George Walling. Steve's sense of his own role was crystal clear: He would provide long-range leadership and direction and let the circle of strong lieutenants he'd gathered worry about day-to-day concerns.

The ability to make bold strategic shifts, particularly when a company is doing well, is the hallmark of visionary corporate leadership, and Steve possessed it. "Moving into engineering was a major change and a very important one," said Bridges. "And Steve got very little support from the rest of us to do it." He pushed equally hard for international diversification. "Nobody around here wanted to go foreign," observed Bill Waste, Steve's senior inside manager. "But Steve kept hammering away and one thing led to another, and we got the job in Venezuela. We didn't have a policy then. Nobody knew what a policy was. We just evolved into these things. He was so far ahead of his own people that sometimes it was hard to understand what he was really thinking about. None of us knew what nuclear meant. It was so far beyond our knowledge we never would have gotten into it. It took vision. And that vision was 99 percent Steve."

Steve and Laura Bechtel. Their close marriage established a Bechtel company tradition of wives accompanying their husbands on most business trips.

He even turned a bad error into a strategic turnaround. A costly mistake on a tunnel project convinced him that Bechtel should concentrate on services such as engineering and construction management, and that they should avoid heavy financial commitments associated with grand projects. His preference for services was reinforced by his aversion to debt, because he wanted to control his own destiny. And Steve Sr. was a masterful salesman—more than any other single person, he was responsible for the company's business development.

In 1960, at the age of 60, Steve decided to step aside and turn the business over to his son, Steve Jr. The elder Bechtel moved up briefly to chairman before assuming the title senior director, a position which, he happily observed, carried with it absolutely no authority and no responsibilities. Although Steve Sr. would remain a trusted adviser, there was no question who was now in charge. For a vigorous 60-year-old CEO who was head of a thriving, worldwide business, giving up the reins was not easy, as Steve Sr. would later concede.

But that, too, is a mark of leadership—knowing the difference between long enough and too long. And it was part of what made Steve Bechtel a salesman without peer, a successful businessman, and the proud builder of a family enterprise that became a world leader in its field.

Eddie Hironaka, The Image Bank

CHAPTER THREE

PEACE AND THE POSTWAR BOOM

Bringing Energy to the World

1945–59

The war's end gave Steve Bechtel an opportunity to do something he had wanted to do for years: relax in an informal retirement. He did so, beginning in late 1945, but not before liquidating most of the company's wartime operations. This also meant parting ways with longtime partner Henry Kaiser, who planned to stick with manufacturing. Bechtel-McCone and W. A. Bechtel Co. were consolidated, and in 1945 a single enterprise was organized in its place, Bechtel Brothers McCone Co. (BBM). The new structure was headed by Bill Waste, a company veteran who became the first nonfamily member to lead the firm.

BBM made a big breakthrough in Steve's absence, grabbing its first piece of major power work, for Southern California Edison (SCE), in 1945. SCE was converting its frequency from 50 cycles to the U.S. standard 60 cycles. Under the brilliant management of John Kiely, BBM was able to whisk the job away from the eastern firms. It was a huge victory and the first of many jobs Bechtel completed for SCE.

The rest of the operations ran the gamut. There was a food plant for H. J. Heinz that could process a ton of tomatoes a minute, and a futuristic flour mill for General Mills. Bechtel pipeliners continued to advance technology. On the Kettleman Hills pipeline in California, they pioneered semiautomatic welding, high-strength pipe, and gamma-ray photographs to spot-check welds. There was heavy activity in chemical plants and petroleum refining. There was a breakwater,

OPPOSITE: *In 1945, Southern California Edison commissioned Bechtel to light up Los Angeles by converting power equipment to the U.S. standard 60-cycle current.*

a canal siphon, a tinplate mill and a steel mill, printing plants, and, in Missouri, the nation's first coal hydrogenation plant, a pilot plant for producing synthetic motor fuel from coal.

Within months, however, Steve was eager to get back into action, and senior Bechtel executives, Waste included, couldn't wait. Recalls Jerry Komes, a vice president at the time, "Bechtel Brothers McCone wasn't doing terribly well. It was viable, but not on the scale the partners were used to. They wanted Steve to come back." Steve formally returned as president at a directors' meeting on October 8, 1946. His statement to BBM's officers was characteristically low-key. "My chosen field has been and is engineering and construction," he told the directors. "It is the most interesting, most stimulating and, in my opinion, one of the finest professional and business fields."

THE BIRTH OF THE MODERN BECHTEL CORPORATION

During his brief "retirement," Steve had thought very carefully about how he would restructure the company and how and where it would operate. Ken, Warren Jr., John McCone, and he agreed that BBM would have to be a more structured entity, run like other modern corporations with divisions overseeing various lines of work. And like other modern corporations, it would be run by professional managers, whether family or not. To accomplish his new goal, Steve formed a series of new corporate structures, a set of spokes with himself at the hub.

The strategic focus of the new Bechtel would be service, working with customers as much as for them. There would be no big capital outlays and no heavy borrowing. The company would travel light and finance its work with payments from customers who paid their bills on time. They would be selective about customers, pursuing only the best.

They also decided to concentrate on the biggest jobs, high-impact work that attracted industry attention, jobs that Bechtel would conceive and design as well as build. The new company would also seek to get half its work from overseas, a balance that would smooth out the inevitable peaks and valleys of the construction business; when one market was slow, another would pick up. Bechtel would diversify into targeted fields and go after projects in those areas as a further hedge against the cyclical swings of construction.

1945

Shortly before V-E Day, W. A. Bechtel Co. breaks ground in Tracy, California, for one of the nation's largest food-processing plants, for the H. J. Heinz Co. The plant is the first major civilian project authorized since wartime regulations took effect. Among the plant's products: baby food for the approaching boom.

1945

After the war, Bechtel Brothers McCone Co. is formed, bringing together the various functions performed by W. A. Bechtel Co. and Bechtel-McCone. Steve, Warren, and Ken Bechtel retire from active management—for now.

1945

BBM signs on to handle equipment changes for Southern California Edison; engineers must convert power equipment operating on a 50-cycle current to a 60-cycle current. The project marks the beginning of the power division.

Steve headed off to Saudi Arabia in January 1947, leaving his senior managers to sort out the details of the new organization. Bechtel had been in the Middle East since 1943, working on a refinery in Bahrain and other projects. Steve's first postwar sales trip produced a bonanza within weeks; word filtered back that Bechtel had contracted to build a major portion of the Trans-Arabian pipeline, called Tapline, linking the oil fields of the Arabian Gulf to the Mediterranean.

BECHTEL IN THE MIDDLE EAST

When a small group of Bechtel workers arrived in eastern Saudi Arabia in July 1947 at what would later be a major base camp for Tapline, they beheld a large expanse of sand bordering the sea. There wasn't even a Bedouin camp to break the monotony. There was no vegetation and no fresh water. The temperature would have been around 120 degrees Fahrenheit in the shade—had there been any shade.

Saudi Arabia in 1947 had a forward-looking monarch, King Abdul Aziz Bin Saud, and a rich culture, but the desert nation had only a rudimentary infrastructure. The king dreamed of a country that could match any in the world with highways, utilities, airports, and the other manifestations of modernity—exactly the things that Bechtel was in the business of building.

Bechtel had worked in the Middle East even before the Saudi king requested its services. In 1943, during the war, Bahrain Petroleum Co. Ltd. (Bapco), a joint venture of Standard Oil Co. of California and the Texas Co., needed to double the capacity of its Bahrain refinery to 65,000 barrels per day and build a fluid catalytic cracker to produce 100-octane aviation fuel. To support and protect that output, Bapco also needed additional storage tanks and a pipeline 33 miles across the Straits of Bahrain, 16.5 miles of it underwater, to the Saudi port at Ras Tanura, a spit of land jutting into the Arabian Gulf. Bechtel was called upon to enlarge the Bahrain refinery, build the cracker, and construct associated facilities. The next year, Bechtel received a contract from the Arabian American Oil Co. (Aramco) to replace a small refinery at Ras Tanura with a modern one. Since that time, Bechtel has undertaken many additional assignments at Ras Tanura. At the writing of this book in 1998, Bechtel is completing a major upgrade of the refinery.

Work at Bahrain began in 1943. In the months that followed, 40,000 tons of structural material flowed from American ports to Bahrain despite constant

1947

Bechtel Corp. is formed, including Compañía Bechtel, S.A., Bechtel International Corp., and International Bechtel, Inc. Steve Bechtel comes out of retirement and takes direct charge of all operations.

1947

Bechtel begins work on the 1,068-mile Trans-Arabian pipeline (Tapline), which will transport oil across Saudi Arabia, Jordan, Syria, and Lebanon to tankers in the Mediterranean. To support the project, Bechtel constructs a power plant to provide electricity, drills wells for water, and builds a road between the Arabian Gulf and Jordan.

The 50,000-barrel-a-day refinery at Ras Tanura was Bechtel's first contract in Saudi Arabia. Completed in 1945, the refinery was expanded in the following years so that by the mid-1950s output had more than quadrupled.

harassment from German warships. Some 750 workers and several hundred Bapco employees were transported to Bahrain to work on the job. Housing, offices, and health and recreational facilities were built. Refrigerated ships supplied food for the workers. The aviation fuel project was completed in 1945, and the program of refineries, piers, pipelines, power plants, and utilities was finished in 1951.

While construction was getting under way in Bahrain, Bechtel began design and engineering work for the refinery at Ras Tanura. Bechtel was building the foundations of a partnership with Saudi Arabia that would last for decades. Steve Bechtel forged solid working relationships with King Saud, his son Prince Faisal, and a tight circle of Saudi advisers. In 1946, Bechtel would begin the task of constructing a modern nation in the Arabian Desert. Over the next few years, it built part of a 350-mile railroad from the Arabian Gulf inland to the capital at Riyadh, a power plant to electrify the capital, a deep-water pier on the Red Sea in

Jiddah, radio stations, and still more power plants. So closely was the company associated with the supply and distribution of electricity that in some newly connected Saudi households, one family member would ask another to "turn on the Bechtel." Later, there would be modern airports at Riyadh and Jiddah, and a new highway linking Jiddah with the holy city of Mecca.

Saudi Arabia's potential for economic growth and development was powered by a seemingly inexhaustible supply of fossil fuels. By 1947, four years after Bechtel first set foot in the Middle East, Saudi Arabian oil production had reached 250,000 barrels a day and was climbing steadily. There was much, much more available, but the country's ability to move its primary source of revenue was limited. That crucial stream would flow through Tapline, the 1,068-mile Trans-Arabian pipeline from the Arabian Gulf to the Mediterranean Sea. And Bechtel would help build more than 800 miles of it.

TAPLINE

Tapline would reroute oil to Europe through the Mediterranean. Up until that time, Middle East oil moved south by tanker through the Arabian Gulf, through the Indian Ocean to the Red Sea, and then north through the privately owned Suez Canal. The 3,500-mile trip could take 12 days and cost oil companies 18 cents for every barrel that passed through the canal, or about $40,000 for each modern tanker. Tapline would deliver oil at a fraction of the cost.

It was a pipeliner's dream. "This 30-inch, 400,000-barrel-per-day line will be the mightiest pipeline ever laid," Steve announced with pride, "bigger than any oil line yet completed and almost as long as the Big Inch line running from Texas to New York." And this, he added, was only the beginning: "I cannot help but foresee tremendous possibilities pointing towards potentially the biggest development of natural resources ever undertaken by American interests."

In July 1947, work commenced. The pipeline began at the rich Abqaiq field, where 18 wells churned out 200,000 barrels of oil every day. Bechtel was responsible for 850 miles of Tapline, from the Arabian Gulf to Jordan. From Jordan, it would continue to the city of Sidon, on Lebanon's coast.

Bechtel field engineers put all their experience to work. Among the many technical innovations was the first over-water use of a cableway "skyhook"—

A water tower was erected in the tent city at Ras al Misha'ab, the base camp for Tapline. When engineers arrived in July 1947, they found plenty of sand but little else. By the end of the year, they had built a small town with its own power and water supply.

Tapline's trucks were specially modified to carry 93-foot lengths of pipe. The loads weighed as much as 40 tons.

borrowed from loggers in America's Pacific Northwest—to transport pipe and supplies from ships anchored in deep water nearly three miles offshore. They put double radiators on all the trucks to cool the engines.

With Tapline, Bechtel and its four joint-venture partners—Morrison-Knudsen, H. C. Price, J. H. Pomeroy, R. A. Conyes, and Sverdrup & Parcel—had constructed the world's biggest pipeline. Tapline also had an unintended, and positive, local impact. Originally, Tapline engineers had anticipated moderate amounts of road-building along the pipeline route. But as work progressed, they had to build a 930-mile highway from Ras al Misha'ab to intersect with an existing road in Jordan. Trucks replaced camels on this new route, allowing merchants to haul fruits, vegetables, and other goods from Mediterranean ports to Arabian Gulf markets. In drilling new water wells, the builders of Tapline created a string of oases that became gathering places for the nomadic Bedouin tribes of the northeastern desert. In both instances, the result was an increase in commercial activity and an improved standard of living.

At times, Tapline fell victim to unrest in the Middle East. Construction was interrupted in 1948 by the first Arab-Israeli war. Syria and Lebanon did not agree to let Tapline proceed until early 1949. But by the end of 1950, oil from Aramco wells was being metered at the Qaisurnah pump station. From there it flowed through Tapline to the new terminus at Sidon, where it was loaded onto tankers for Europe. The geography and economics of oil had been completely altered. Suliman Olayan, a government liaison officer for Aramco who had earlier met King Abdul Aziz and Steve Bechtel, supplied two trucks and labor services to Tapline and went on to become one of the world's most successful businessmen.

The process of development would also bring with it one of the most significant transfers of technology and training ever seen. Bechtel's prime contribution to most of these projects was experienced guidance. The bulk of the work was done by nationals employed through Arab subcontractors. And as the work progressed and the subcontractors gained experience, their role and influence grew.

THE POSTWAR ENERGY BOOM

For suppliers of anything from baby carriages to power plants, the 1950s was an age of nearly unrelenting demand. And just as Steve Bechtel had predicted, the demand for energy of all kinds grew steadily through the decade. America's love affair with the automobile was rekindled, and Congress obliged by lacing the nation with interstate highways. In Europe, the demand for petroleum products would double as the continent shifted from coal to oil. And developing nations around the world began tapping their own rich oil reserves in response.

Bechtel launched a series of major projects that gave the organization the diversified look that would characterize it for years to come. Overseas, the company was building power plants in South Korea, rail tunnels in New Zealand, and oil pipelines across Iraq and Syria for the Iraq Petroleum Co. and through the jungles of Sumatra for Caltex. Increasing demand for oil kept Bechtel busy expanding projects in Saudi Arabia and Kuwait.

North America, too, saw its share of signature projects, many in Canada, where Canadian Bechtel Limited was established in 1949. These included a 439-mile

Arab workers quarried material for a railroad pier in 1948. The railroad, one of the many pieces of infrastructure necessary for modern transportation, was crucial to Saudi Arabia's development.

segment of the Interprovincial pipeline, linking the oil fields of Alberta with the Great Lakes; the 643-mile Lakehead pipeline extension; and, in 1953, the world's deepest underwater pipeline, as much as 238 feet beneath the Straits of Mackinac. Management and construction of major natural gas pipelines included the 2,300-mile Trans-Canada system serving Ontario and Quebec, the 800-mile Westcoast Transmission line supplying British Columbia, and the Tennessee Gas Transmission lines serving much of the eastern and midwestern United States. There were power plants in California and Hawaii, a nuclear power plant in New York State, and hydroelectric plants in Washington State.

The Bechtel organization had to expand and reconfigure itself to better respond to constantly shifting markets. An International Division was formed, the Power & Industrial Division was created, and the new Scientific and Nuclear

The Kuwait oil pier and marine terminal was completed in 1949. Servicing as many as eight tankers and handling up to 1.2 million barrels of oil per day, it was the largest such facility of its time.

Development Department came into being, as did Arabian Bechtel Corp., Bechtel Méditerranéenne, and International Bechtel Builders, Inc.

Bechtel was also expanding in other fields during this decade, including building its first aluminum plant and providing project management services to Inter·Continental Hotels for hotels worldwide. There was the Orinoco iron ore project, begun in 1951, deep in the interior of Venezuela. It took three years to open this new source of high-grade iron ore, the largest facility of its kind in South America.

THE TRANS MOUNTAIN PIPELINE

One of Bechtel's most dramatic projects on the North American continent was the $93 million, 700-mile Trans Mountain pipeline. Begun in 1952, it was built to carry crude oil from Alberta and, after climbing to an elevation of some 3,600 feet across the Canadian Rockies, deliver it to the Pacific Northwest. It was a grueling job. Construction crews ran into formidable natural barriers: rock-walled canyons, cascading rivers, and slide-prone cliffs. Bechtel was concerned, too, that the pipeline not disrupt the environment, a concern that further complicated the construction process. The first delivery of Alberta crude was made in Vancouver in the fall of 1953. Impressive as it was, the physical accomplishment of completing this sweeping high-rise pipeline was just as important for the organizational structure that made it possible.

The Trans Mountain Oil Pipe Line Co. was formed with Steve Bechtel Sr. as its first board chairman and Robert Bridges as president. It was the first major engineering management job in which Bechtel functioned as the owner's representative with overall responsibility for design, engineering, procurement, and management of construction. In addition, Bechtel inaugurated a new concept in management, with the company handling the financing, making the executive decisions, and developing an operating organization. On completion of Trans Mountain in 1953, Interprovincial hired Bechtel to manage the engineering and construction for the Lakehead pipeline extension. "With these two successive management jobs," Bridges observed, "Bechtel had a prototype which it used in successive contracts, including most of the pipelines in Europe and North America."

Bechtel is chosen to engineer and build a $25 million manufacturing plant in Los Angeles for Lever Brothers, the largest industrial investment by private business in the history of the area. The plant manufactures soap, shortening, and salad dressing.

Just as Bechtel had avidly pursued joint ventures in an earlier generation, so it now took the lead to establish itself as manager of virtually every phase of a project from conception to completion. "We took an overall role as the right arm of the owner," said then–Vice President Jerry Komes. "We developed a technique of management that allowed the client to run his own business while we took care of his construction. The technique proved handy in international settings where national pride can be a touchy issue. The Europeans and Japanese felt they were quite capable of building, so Bechtel could assume an overall position and leave the construction to the locals. We developed more flexibility, which opened up more opportunities."

To help frame solutions, Steve Bechtel would hopscotch the world, dropping in on heads of state and international businessmen to chat about their concerns.

Just after the Korean War, Steve visited South Korea and met with President Syngman Rhee, who complained of the problems he was having getting the American government to help rebuild his war-torn country. The president told Steve that he needed roads, factories, power plants, everything. Steve suggested he build identical power plants and use Korea's copious supply of cheap coal for power. In 1956, the Masan, Tangin-Ri, and Samchok power plants went on-line, doubling Korea's energy output.

During this period, according to Komes, "Steve operated on a more relaxed basis. He would take a trip around the world with no particular target in sight." Komes and Bechtel would fly to London for lunch with old friends from British Petroleum or to pay a courtesy call on the head of Imperial Chemical Industries. In Paris, they would discover that J. Paul Getty was staying in the same hotel, Steve would give him a call, and they would get together to talk about world business— Getty's concession in Kuwait, for example.

"We went around to Beirut, and through Saudi Arabia and Kuwait and then over to India and to Tokyo and the Philippines," said Komes. "[Steve] had an intense drive to be moving and active all the time. He was very sensitive to high-level contacts and he was constantly working them."

When Bechtel moved into Europe, said Bridges, "We were really unknown. People would ask us, 'Beshtel? Who is this Beshtel that you call yourself?'" It didn't

OPPOSITE: *Until Bechtel tackled the job, no one had ever placed a pipeline over the Rockies. Pipeliners on the Trans Mountain project faced a series of streams, cliffs prone to slides, and other natural hazards.*

1951

Near Arco, Idaho, Experimental Breeder Reactor-1 (EBR-1) is the first nuclear reactor to generate electricity from atomic energy.

1952

Bechtel begins construction on the 700-mile Trans Mountain pipeline, which will transport oil from the oil fields in Alberta over the Canadian Rockies to the Pacific Northwest. Trans Mountain is the first job in which Bechtel is the owner's representative, responsible for all the design, engineering, procurement, and management of construction.

1954

Bechtel begins to build three power plants for South Korea, doubling that country's energy output.

take them long to find out. The Pipeline Division welded its way across continental Europe and Saudi Arabia, working out of a London office that had been opened to handle the Aden refinery, which was completed in 1954. And later, when oil was discovered in the North Sea, Bechtel's London presence helped it secure contracts for the first North Sea offshore oil production facility for Hamilton Brothers. In the late 1960s, Bechtel won the Prestige de la France diploma, one of the highest honors awarded industrial firms operating in France. By then, questions about who this "Beshtel" was had long since been answered.

In 1959, Steve produced, along with Morrison-Knudsen and Brown & Root, a study for a tunnel under the English Channel. Three decades later, Eurotunnel, an Anglo-French company, built the Channel Tunnel, which now carries high-speed trains from London to Paris in three hours. Bechtel assisted Eurotunnel during the later stages of its ambitious program, and in 1996 a Bechtel-led consortium won the concession to design, build, and operate the high-speed Channel Tunnel Rail Link in England.

THE NUCLEAR AGE

If W. A. Bechtel's breathtaking risk on Hoover Dam set the tone in the 1930s, Steve Sr.'s, and later Steve Jr.'s, commitment to nuclear power set the tone for the postwar decades and continued a family tradition. The Bechtels would bet the company on their strategic vision. Steve Sr. had been right about energy trends before, foreseeing the oil boom of the 1930s and steering the company into pipeline and refinery construction. Soon after World War II ended, he became convinced that nuclear energy would revolutionize electric power generation, and he wanted to be part of it.

A decade later, nobody doubted that nuclear energy could work. The real question was, Could anyone make a profit in it? Steve Jr., who would become president of Bechtel in 1960, proved it possible, but only after making an exponentially higher wager than his father had made on pipelines. Steve Jr. also learned to work with such giants as General Electric and Westinghouse in the process.

In late 1949, the company got an early introduction to nuclear energy as a government contractor. Engineers from Bechtel's refinery division helped design the Atomic Energy Commission's Van de Graaff nuclear accelerator at Los

UPI/ Corbis-Bettmann

Alamos, New Mexico, in 1948, which led to a contract to build the AEC's Experimental Breeder Reactor-1 (EBR-1) in late 1949. Set on a lava and sagebrush plain at an old naval proving ground in Idaho, EBR-1, a modest 100-kilowatt plant fueled by uranium 235, became, on December 21, 1951, the first reactor to generate electrical power from nuclear fission.

When Congress passed the Atomic Energy Commission Act of 1954, allowing private companies the right to build and operate nuclear power plants, the scramble for contracts was on. At Steve's urging, the Bechtel board agreed to an aggressive "Triple Ten" strategy. Bechtel would commit 10 percent of its pretax profits and 10 percent of its management and engineering capability for 10 years to learning the technology of nuclear power. "It was an investment in technology," said Steve Sr., "in positioning ourselves in an industry that we were convinced was going to be very big."

Bechtel had joined with Pacific Gas and Electric and six major eastern and midwestern utilities to form the Nuclear Power Group (NPG), a Chicago-based

The Korean War left that country devastated, with only 70 megawatts of electrical power for a population of 22 million. Candles and oil lamps were used for home lighting.

1957

Groundbreaking for the Dresden nuclear plant takes place in Illinois. Bechtel helps design and construct the plant, which is completed in 1959.

1958

Bechtel begins a long relationship with the Inter•Continental Hotels unit of Pan American World Airways. Bechtel assists with site selection, materials and equipment, and construction planning, as well as construction management for 22 luxury hotels over the next 10 years. The first four hotels are in Beirut, Vienna, Melbourne, and Geneva.

trade organization that promoted nuclear power and undertook economic and design studies for the AEC. By 1955, the group had completed a functional design for a nuclear plant; to build it, the group would have to incorporate. Steve Bechtel Sr., Perry Yates, and Robert Bridges flew to Chicago to discuss Bechtel's involvement with representatives of the consortium. Willis Gale of Chicago's Commonwealth Edison laid out the deal. The NPG would incorporate and commit $15 million to research and development of its first nuclear plant. The utilities would each put up $2 million or $3 million. Bechtel would have to commit $1 million cash. It was an immense sum for Bechtel, but Steve decided that the company could not afford to miss the opportunity.

In 1955, the NPG went to the AEC with plans for Dresden-1, a 180-megawatt boiling water reactor, the world's first large, privately financed, all-nuclear commercial power station. Commonwealth Edison put up $30 million for construction costs. Bechtel and GE had worked closely on the design, and GE now offered to fix the entire plant cost at $45 million if Bechtel would fix its portion at $33 million.

Having put up $1 million to join the nuclear club, Bechtel was now being asked to risk millions more to exercise its membership. Steve presented the proposal to a group of wary directors, some of whom had strong doubts about risking so much on an uncertain technology. He argued forcefully that, risky as it was, Bechtel had to go ahead, that the company had to be at the forefront of this revolutionary new technology. "He didn't try to bulldoze anybody," recalled John Kiely, a young director at the time, "but he obviously wanted to do it."

Steve got his way. Construction on Dresden began in early 1957 and was completed in 1959. "Dresden did more to establish commercial nuclear power than any other single project," Steve Bechtel has said. "It set the precedent for the utilization of an independent engineer-constructor and established the roles of the owner, engineer-constructor, and manufacturer in this type of work. It justified public utility boards in authorizing additional nuclear plants." During Dresden's construction, Bechtel won fixed-price contracts to build plants for PG&E at Humboldt Bay in California and Consumers Power of Michigan at Big Rock Point.

Bechtel soon won a fixed-price contract for Peach Bottom I in Pennsylvania, for Philadelphia Electric. Peach Bottom I proved to be an

extremely costly lesson in dealing with untested technology. Problems with the new gas reactor caused significant delays. But Bechtel pressed ahead and broke some important ground. Peach Bottom was the first nuclear power station to use a graphite-moderated, helium-cooled reactor, making it one of the most advanced nuclear plants of its time. Careful attention was paid to environmental controls. Water was used to condense steam from the plant's turbines, and recirculated through natural-draft cooling towers, instead of being discharged into the river. The company later profited from its expensive ride up the learning curve, and went on to build two more plants at Peach Bottom and two at Limerick, Pennsylvania, for Philadelphia Electric. It was a lesson Steve Bechtel Jr. would bear in mind later.

The Swift hydroelectric project on the Lewis River in Washington, completed in 1958, was at that time the highest earth-filled dam in the world, rising 512 feet above its foundation.

1959

Work begins on the Bay Area Rapid Transit (BART) system in the San Francisco area. BART is the first of many mass-transit projects for Bechtel worldwide. In 1976, it becomes the first totally new rapid transit system to be completed in the United States in 40 years.

CHAPTER FOUR

STRUCTURING FOR A NEW ERA

Advanced Technologies and Modern Management

1960–69

As Bechtel's sixth decade came to a close in 1958, the organization had compiled an impressive record, having worked on 2,000 projects in 40 states and in 30 countries on six continents. But in many ways, Steve Bechtel Sr.'s most important creation was not a physical structure that transformed some remote corner of the world; it was the organization he built and the attitude he brought to its work. In 1960 at the age of 60, he decided it was time to turn the reins over to the next generation, to pass along the responsibility to his son, Steve Jr.

Stephen D. Bechtel Jr. was just 35 years old, a child by corporate standards, but he became president of Bechtel in a time that seemed to call for new, young leaders. Steve Jr. was ideally suited in terms of education, experience, and temperament to oversee Bechtel's corporate transition to this new age of organizational complexity.

The nation was just recovering from a modest recession—a mere breather after the torrid postwar growth phase—and about to embark on a prolonged period of robust economic expansion. In 1960, Bechtel's revenues were more than double those of 1955 and four times the 1950 figure. The company's salaried workforce had climbed to an all-time high of nearly 4,000.

OPPOSITE: *Bay Area Rapid Transit (BART) system trains glide on elevated tracks. As BART shuttled San Francisco Bay Area residents to work and play, its use of computerization and automation raised mass transit to a new level of sophistication.*

Athabasca Tar Sands project is developed in northern Alberta for Great Canadian Oil Sands Ltd. (subsequently Suncor). It is the first large-scale operation to recover oil from the area's enormous tar sand deposits. The plant uses hot water and steam to separate the bitumen from the sand.

NASA engineers examined the Gemini Mission Simulator, which duplicates the orbiting spacecraft in configuration and operation. Bechtel crews played a key role in supporting Gemini and Apollo space projects.

Energy use, fed by growing economies everywhere, was on the rise, fueling strong demand for petroleum products, natural gas, and electric power. The need for production, processing, and transportation facilities was increasing dramatically. New projects were getting bigger and more venturesome. This was also the golden age of space flight; anything was possible. Bechtel's markets never looked better, and Bechtel was in an excellent position to respond.

But the 1960s would also prove to be an extremely stormy decade. The civil rights movement, the antiwar movement, and environmental activism would change the country forever. Corporations would be no exception.

In hindsight, Steve Jr.'s strengths seem well matched to the demands of this decade of sharp contrasts. For one, he shared his father's instincts for opportunities. He was willing to make big bets on his thoroughly researched sense of where industrial markets were heading. At the same time, the people of Bechtel needed strong management; they would get it with Steve Jr.

MAINTAINING AN ENTREPRENEURIAL SPIRIT

As a sort of welcoming present to his son, Steve Sr. commissioned another soon-to-retire executive, Finance Chairman John L. Simpson, "to review our past policies, philosophies, and practices and give me your recommendations on how they might logically be changed for the present and the future." He wanted Simpson to articulate what made Bechtel successful, and what it would have to do to continue on that road.

Leadership, said Simpson, had been the key to Bechtel's achievements. Steve Jr. would have to preserve as much of that as possible to maintain an entrepreneurial spirit and sense of individual involvement and responsibility. Simpson also encouraged him to maintain Bechtel's flexibility, its readiness to deal with new developments.

But because of Bechtel's growth, Steve Jr. would need to create a more sophisticated, tightly woven management structure that could manage complexity and risk as well as provide some sort of continuity in a business that had historically been characterized by an endlessly repeated series of beginnings and endings. The relentlessness of that cycle had always dictated an emphasis on immediate accomplishment rather than long-range planning and personnel development. But

business now was less influenced by a handful of key customers. The number of major projects had increased to the point where no particular customers dominated the scene. This represented a dramatic shift from Steve Sr.'s earlier determination to concentrate on large projects for a small, select group of major customers.

Report in hand, Steve Jr. began working on a major overhaul. His arrival marked a generational change throughout the company, a gradual shift of management functions to a seasoned team of younger executives. He simplified the organizational structure, clarified the lines of responsibility, and strengthened top management, which already boasted such leaders as Bill Waste, Perry Yates, John Kiely, and Jerry Komes.

INTERNATIONAL PATHBREAKING PROJECTS

While the management of Bechtel was evolving, the building of everything from pipelines to energy plants continued unabated throughout the world. Bechtel entered the 1960s with a wealth of landmark international projects:

At the Palabora copper mine, in addition to building ore-crushing plants, service buildings, a smelter, a power plant, and a laboratory, Bechtel helped create a storage dam, a railroad, and miles of underground pipelines.

The Alberta-California pipeline was completed in 1961. Stretching for 1,400 miles, it brings Canada's rich supplies of natural gas to the Pacific Northwest and California.

— Western Europe's first international crude oil pipeline, the Rotterdam-Rhine pipeline, was completed by Bechtel in June 1960. Built at a time when Europe was converting from a coal to a petroleum economy, the pipeline was a significant example of the increasingly integrated European community.

— Bechtel served as engineer-manager on the Texaco Pembroke refinery in the United Kingdom, incorporating some of the most advanced automation control equipment in use at the time.

— Bechtel built the Republic of Panama's first oil refinery in 1961.

— With Bechtel's completion of a saltwater conversion plant on St. Thomas in 1961, tourist boards rejoiced, since residents and visitors to the Virgin Islands no longer had to obtain drinking water from Puerto Rico or from basins that caught rainfall. Every day, the new plant supplied 275,000 gallons of fresh drinking water from the Caribbean Sea.

— Bechtel oversaw the engineering, procurement, and construction of the largest chemical plant complex ever built at one time, the eleven-plant Ponce petrochemical facility in Puerto Rico, completed in 1972.

– Bechtel built the first long-distance iron-ore slurry line in the world, the 53-mile Savage River project in Tasmania, completed in 1967. Construction required throwing a 1,200-foot suspension bridge over the nearly inaccessible Savage River. This 9-inch slurry pipeline was a pioneering method of transporting iron ore and other solids. The ore was crushed at the mine site, then flushed through the pipeline with water to its final destination. There it was reconstituted into a solid.

– One of the most modern petrochemical projects of the time, the Chocolate Bayou plant in Texas was built by Bechtel and featured on-line digital computer controls. Detailed engineering began in November 1960, ground was broken in 1961, and all units were turned over to the customer, Monsanto, in September 1962.

1962

The Bechtel-built Chocolate Bayou petrochemical plant is dedicated in Texas. It is one of the largest ever built, with all the units constructed at the same time. The plant produces chemicals used in packaging, flexible pipe, floor tile, and electrical insulation.

THE TRANS-ALPINE PIPELINE

Although Bechtel was diversifying into new arenas, pipelines continued to be big business. One of the most challenging projects was the Trans-Alpine pipeline, a system that traversed parts of Italy, Austria, and Germany. When it was completed in 1967, Trans-Alpine carried 500,000 barrels of crude oil per day. Bechtel was responsible for the feasibility study, procurement, project engineering, and construction management of nearly 300 miles of 40-inch pipe, from the Gulf of Trieste over the Alps to central Bavaria. The pipeline, which reaches mountain-goat elevations, required three four-mile tunnels bored into the Alps and five pumping stations.

Bechtel pipeliners built the 1,400-mile Alberta-California natural gas pipeline, which was completed in 1961, then expanded in 1965 and again in the 1990s. Planning for this immense project began in 1956. The project enhanced Bechtel's growing reputation for managing large, complex undertakings.

Another European pipeline was begun in 1961, the 470-mile South European pipeline. The route made a total of 2,500 crossings, including the Durance, Rhône, Isère, and Drôme rivers; roads; canals; and railroads. The longest crude oil pipeline in Europe, it shortened the total tanker-pipeline transit time from North Africa to Central Europe by five days, lowered total transportation costs, and made possible a major refining center in the Alsace–Upper Rhône area.

The company faced one of its stiffest management challenges in nuclear energy, a field that was growing rapidly and attracting many competitors. From 1960 to 1965, Bechtel watched its early lead in nuclear power slip away. The building of nuclear power plants was a dazzling technological feat in the 1950s, but it wasn't long before a number of companies could do it. To complicate matters, turbine and generator manufacturers such as GE and Westinghouse, having made huge investments in nuclear energy, were now keen to protect their business. Bechtel found itself in an escalating battle with these competitors, leading to a showdown at Turkey Point, a nuclear facility for Florida Power & Light (FPL) to be located in a mangrove swamp south of Miami. Bechtel had already built two oil-fired electric plants at the site for FPL, and Steve Jr. fought hard to win the project.

Westinghouse, GE, and other manufacturers were pressing clients to make them the prime contractors on such turnkey jobs, thus ceding them control over all aspects of the work. Not Bechtel. "We didn't want to make reactors," said Harvey Brush, then engineering boss of the Power & Industrial Division and later executive vice president of Bechtel Group, Inc. "But we didn't want them to do engineering and construction." Despite its early leadership, by the 1960s Bechtel found itself working largely as a subcontractor. To head off the other old-line builders, Harry Reinsch, who was responsible for many of Bechtel's power plant developments, offered an unprecedented multiproject contract, on a lump-sum basis.

But there was a problem, and Reinsch knew it—the deal would lose money for Bechtel, and he couldn't bear it. He walked away from the deal, telling the FPL boss, "We just can't do it." Back in San Francisco, a stunned John Kiely, who was senior vice president at the time, was as emphatic as Reinsch was uncertain. "Harry," he told Reinsch, "you've gotta go back and get that job." Kiely, too, knew that Turkey Point was a loser, but he argued that making it a loss leader would guarantee Bechtel a future in the nuclear business. If Bechtel lost this job, said Kiely, it would never be a major player in the nuclear industry.

Bechtel would take a $20 million loss on the two nuclear plants, Turkey Point Three and Four. But it brought in the Florida plants at a lower cost to its clients than the reactor manufacturers ever had. Bechtel had lost money, but it made the industry a viable sector for work again.

By 1968, three years after winning the battle for Turkey Point, Bechtel had also completed the San Onofre nuclear power plant for Southern California Edison, which, at double the size of any existing U.S. plant, was then the largest single-purpose nuclear power plant in the United States. At this point, Bechtel had completed or was at work on 27 nuclear-fueled generating units and had built up a backlog of work that would last more than 15 years. "[Turkey Point] was the domino that had a whole row behind it," said Brush. "Once it fell, the whole chain went down—Palisades for Consumers Power, Point Beach for Wisconsin Electric, Monticello for Northern States Power, Calvert Cliffs for Baltimore Gas & Electric, and Peach Bottom II and III for Philadelphia Electric." Bechtel's dominance was unquestioned. By the mid-1980s, the company was responsible for 40 percent of all nuclear work in the United States and half the nuclear plants in developing countries.

Though Bechtel sacrificed financially to win the contracts to build the Turkey Point nuclear power plants in Florida, completing them successfully and economically established the company's preeminence in the industry.

SCIENTIFIC DEVELOPMENT AND NEW TECHNOLOGIES

The nuclear experience proved not only that demands for new technology surface all the time, but also that shifts occur quickly. To stay on top of these opportunities

1968

Bechtel completes work on the San Onofre nuclear power plant in California, which is double the size of any other nuclear plant in the country, generating up to 450,000 kilowatts—enough electric power for a city of half-a-million people.

and challenges, the company's Scientific and Nuclear Development Department was expanded. The department conducted applied research and development and made work for the U.S. space program possible. The department was also instrumental in the development of the Argonne fast breeder reactor, a type of nuclear reactor that manufactures plutonium as a by-product. Research into the application of microwaves to industrial processes resulted in the design and construction of the first non-Bell microwave communications system.

Typical of the application of new technologies was the Great Canadian Oil Sands (later named Suncor) project at the Athabasca tar sands, built between 1962 and 1967 in northern Alberta. Here, Bechtel helped develop the first successful plant to process tar sands into high-grade synthetic crude by the use of hot-water separation. The tar sands in this remote region, 600 miles south of the Arctic Circle, contained twice as much oil as all the world's known conventional petroleum reserves, but until then no one had figured out an economical way to separate the oil from the sand. Bechtel developed a system whereby giant bucket-wheel excavators scooped up the tar sand onto high-speed conveyors, each with a capacity of 100,000 tons per day, and carried the sand to the first step in a process that produced high-grade synthetic crude. When the project, including a 266-mile pipeline, was completed, Suncor would produce 45,000 barrels a day of synthetic crude. The Athabasca project demonstrated how designers, engineers, estimators, constructors, and procurement specialists could collaborate in a common, integrated effort.

MINING AND METALS

Bechtel's expertise in transforming oil and electric power and its experience working in isolated or inaccessible regions were further utilized in its work with metal resources. The 1960s brought a new era for the metals industries, resulting from new recovery processes, many of them enhanced by Bechtel engineers. The Carol Lake iron ore beneficiation project in the remote lake region of western Labrador, 600 miles northeast of Montreal, was one of the largest beneficiation plants in the world. So remote was its location that the project initially required development of a town site, airport, and railroad. In the end, it processed more than 15 million tons of ore annually.

The extraction processes used by Bechtel at Athabasca did double duty in the company's extensive mining and metals projects in the Southern Hemisphere. In South Africa, from 1963 to 1972, the company worked on the Palabora copper project, one of the world's largest open-pit copper mines. Designed by Bechtel, Palabora also used the world's largest grinding mills. The original smelter produced 80,000 tons of 99.4 percent pure copper anodes, used ultimately in the manufacture of copper wire and cable.

Another Bechtel-sponsored joint venture, the Bougainville copper project in Papua New Guinea, was one of the world's largest grassroots copper mining and processing facilities. The project is noted for implementation of one of the most complex telecommunications networks linking a remote job site with civilization, demonstrating that no job site is too far away for state-of-the-art communications systems. Construction of a pipeline and access roads through mountainous jungle terrain from sea level to the mine at 2,500 feet was also a major achievement. Bechtel commenced building Bougainville in 1969 and completed it in 1972.

1968

As part of a joint venture, Bechtel signs on to develop the West Irian copper project on the Indonesian part of the island of New Guinea. The copper, extracted from a deposit at an elevation of 11,000 feet, is transported to the coast by a 70-mile slurry pipeline.

Bechtel workers, equipment, and supplies were airlifted into the remote jungles of West Irian in order to build the copper mine there.

Inter-Continental Hotels and Resorts

The Hotel Inter•Continental Paris was restored to its former glory and completely modernized by Bechtel between 1968 and 1970.

A few miles from the highest peak in Irian Jaya in Indonesia, at an elevation of 11,000 feet, Bechtel demonstrated with the West Irian copper project (started in 1968 and completed in 1973) that, once again, no body of ore was too remote or inaccessible. Helicopter transport was used extensively during construction, and access from the mine site to the concentrator, 2,300 feet below, was provided by three aerial tramways.

BUILDING HOTELS AROUND THE GLOBE

In the 1960s, Bechtel also became a premier hotel builder. In the late 1950s, Pan American World Airways Chairman Juan Trippe had urged Steve Sr. to take over the construction of several hotels that the airline's Inter•Continental Hotels subsidiary had under way. With construction hopelessly behind schedule, Steve Sr. at first declined. But Trippe persisted and, as a teaser, sent over a list of countries where new Inter•Continental Hotels were to be built. Trippe's project list included 10 countries that neither Steve Jr. nor his well-traveled father had ever visited, much less worked in. Steve Sr. finally concluded that doing these hotels might provide Bechtel a valuable entrée to a number of newly emerging nations, possibly leading to bigger projects. Bechtel would, father and son agreed, take on two of the hotels and see how they turned out. If things went well, they would do more.

Things turned out very well, indeed. In Zambia, Bechtel's Inter·Continental Hotel was the first of its kind ever completed on time and under budget in that country. Mightily impressed, Zambian officials asked Bechtel to build a new airport. And then a refinery. In the years that followed, Bechtel would build more than $250 million worth of facilities for the government of Zambia. That experience was repeated in Sri Lanka, Zaire, Romania, and a dozen other countries where Pan Am furnished an introduction and Bechtel's work attracted the attention of local officials.

The hotel building program that was begun in 1958 grew rapidly during the early 1960s. By 1967, 22 hotels had been completed or were under construction in 14 countries as a result of a Bechtel-Inter·Continental arrangement.

LIQUEFIED NATURAL GAS

Beginning in the 1950s, Bechtel began to expand its market development studies. From time to time, both Steve Jr. and Steve Sr. had launched research to investigate their often inspired hunches rather than waiting for work, since bringing well-conceived projects to potential customers could create business.

Bechtel's researchers were keenly aware that energy supplies were critical to industry. But which countries needed them most? And who had energy to spare? And how do you connect the two? The answers to these questions led Bechtel to liquefied natural gas (LNG)—natural gas held in a liquid state by lowering the temperature to minus 260 degrees Fahrenheit. Japan needed energy, and Alaska had a plentiful supply of natural gas. A pipeline was impractical, but from 1966 to 1970, Bechtel designed and built for Phillips/Marathon the only plant to export LNG from North America. The facility, built on Alaska's Kenai Peninsula with Phillips "cascade" technology, was a particular challenge because of the severe, cold climate and strong tides. The LNG was transported by ship to Tokyo Electric Power Co. Twenty-five years later, Bechtel and Phillips teamed up to optimize the cascade process. In 1996, after three years of work, Bechtel began building the technology into a new LNG plant in Trinidad for customer Atlantic LNG.

In Papua New Guinea, Bechtel begins work on the Bougainville copper project, which starts production in 1972. Bechtel is responsible for the construction of a pipeline and roads through a mountainous jungle—from sea level to 2,500 feet—as well as the implementation of one of the most complex telecommunications networks to date, linking the remote job site to the outside world.

An offshore tanker loading facility in Alaska's Cook Inlet was built by Bechtel and others. Afterward, large ships no longer had to wait for the four-foot ice barrier around the oil-rich inlet to melt. Ships now had access to Alaskan crude year-round.

OPPOSITE: *The binocular-shaped tube sections of the 3.6-mile-long Bay Area Rapid Transit system's trans-bay tube were fabricated at Bethlehem Steel Shipyard in San Francisco.*

RAPID TRANSIT

San Francisco Bay Area planners knew that something had to be done to relieve the increasing traffic congestion in the region if it were to continue to enjoy growth and maintain its livability. In 1951, California's state legislature created a commission to study the problem. In 1959, the Bay Area Rapid Transit (BART) District authorized a joint venture of Parsons Brinckerhoff, Tudor Engineering, and Bechtel to develop engineering data, a preliminary design, and estimates for the rapid transit system. Three years later, the plans were approved and the District gave the joint venture the go-ahead to perform detailed engineering and manage construction.

The idea was simple—make mass transit an attractive alternative to driving by creating the most comfortable transportation system possible. The execution was not simple—BART's history is as long and torturous as any civic project's.

From an engineering standpoint, the BART project was magnificent. It

was the largest and most advanced rapid-transit project ever undertaken—a 71.5-mile, double-track line that passed through 3 counties and 14 cities with the key link of a 3.6-mile-long tunnel underneath San Francisco Bay. The system has since been extended.

The bay tunnel was constructed piece by piece in 57 sections, each of which was floated into position, lowered into a trench at the bottom of the bay, connected, then covered. Meanwhile, below San Francisco's Market Street, building subway stations 80 to 100 feet underground posed enormous difficulties, and they had to be built under compressed-air conditions to hold back mud and bay water. The subway excavations also turned up buried ships and other artifacts from 19th-century San Francisco. The first segments commenced service in 1972.

FROM TRAGEDY TO TRIUMPH AT CHURCHILL FALLS

In 1534, French explorer Jacques Cartier described Labrador as "the land God gave to Cain." "This country," he noted, "has nothing of any use to mankind." Barren year-round and locked in ice more than six months of the year, this remote region possessed huge amounts of untapped energy in the form of water that poured over Churchill Falls in the vast plateau of central Labrador.

In the early 1960s, harnessing that energy became the mission of Churchill Falls Labrador Corp. (CFLCo). It became Bechtel's mission, too, after the company was chosen, along with Acres Consulting Services, to undertake the development of a hydroelectric plant. The challenge was daunting. The Acres Canadian Bechtel (ACB) joint-venture team would have to devise and build the infrastructure to handle trillions of gallons of water and channel it through an enormous underground power station. The cost would be just under a billion dollars.

Churchill Falls was the largest single-site hydroelectric project ever undertaken. And it was plagued by problems from the start. There was almost no surface access to the remote reaches of Labrador. Machinery and equipment had to be moved by ship from Montreal, Quebec, and St. John's, Newfoundland, to Sept-Iles on the St. Lawrence River, hauled by rail another 287 miles to Esker, and trucked the final 105 miles to the job site on a road hacked out of the wilderness.

Further, tragedy struck on November 11, 1969. A corporate jet belonging to CFLCo flew into a hill while approaching Wabush Airport on a heavily overcast day.

Bechtel's Fred Ressegieu, general manager of ACB, and his two top construction men, Herb Jackson and Arthur Cantle, were killed instantly, as were Donald J. McParland, CFLCo's president, his two senior officers, and the plane's two pilots.

While coping with the human losses, new project manager Steve White and his crew completed the first turbine five months ahead of schedule. Over the next several years, ACB would build and build. Forty miles of dikes. Eleven penstocks, each more than a thousand feet tall. They carved out of solid rock the world's largest underground powerhouse, 972 feet long and 154 feet high. They built 300 miles of new roads, 6 huge concrete spillways, and one of the world's largest switchyards. At the peak of construction, the project employed a workforce of 6,200. With Churchill Falls, Bechtel had built a single-site power system that dwarfed anything in the western world, producing 5,225 megawatts. "It was," said Steve Jr., "a fabulous job of civil engineering."

For Bechtel, the 1960s had been a time of unparalleled growth at home and abroad. The organization's professional staff had more than tripled, reaching around 14,000. As the decade closed, the company was working on about 100 major projects in 60 countries.

The Churchill Falls hydro-electric development was the largest in North America when it was built. More than 5 million tons of rock were excavated to create the development's powerhouse shell alone. Churchill Falls increased Canada's electricity output by 20 percent.

STEPHEN D. BECHTEL JR.

If he had wanted to work at something entirely outside the family business, it's likely that Stephen D. Bechtel Jr.'s parents would have supported him no matter what his choice. "We always told our children that whatever they did, they should do it well," said Steve Sr. "Don't do it unless you're really going to work at it." But the fact was that Steve Bechtel Jr. was made for a career in the construction business. And that's where he was headed.

Steve Jr. began his career with Bechtel in the field, pipelining alongside some of the company's most respected veterans.

Steve Jr.'s earliest memories were shaped by the awesome power and scope of that massive dam in the Nevada desert begun back in 1931. He loved to accompany his father and grandfather on their inspection trips to Hoover Dam. At age 15, he landed his first regular job as a sweeper for the Lorimer Diesel Engine Co. in downtown Oakland. With war approaching, Steve Jr. spent the summer before his senior year in high school working at Marinship. In 1943, he enrolled in the University of Colorado as part of the U.S. Marine Corps Officer Cadet Unit. But he transferred to Purdue, where he completed a four-year engineering program in two and two-thirds years. Later, he graduated from Stanford's two-year MBA program in just 17 months.

Not until he had finished his graduate work at Stanford in 1948, though, did he decide to join the family business. His first position was in the field, working for Perry Yates, one of his father's most trusted colleagues, on a pipeline project in Texas. Steve Jr. then moved from project to project around the U.S. and Canada with his wife, Betty, and their children. On every project to which his son was assigned, Steve Sr. made sure that he worked with the best people, including such Bechtel legends as Yates, Heinie Hindmarsh, and Van Rosendahl, people who could impart technical knowledge and institutional wisdom.

Steve Jr. was well aware of the potential pitfalls in being the boss's son, and he avoided them by working harder than anyone else. As his uncle Ken observed, "Steve was born and raised by his mother and father for this wonderful opportunity—deliberately, successfully, lovingly, from infancy to the time he took over. But no matter how well prepared he was, if he hadn't done his share and more, it wouldn't have happened."

Steve Jr. was just 35 in 1960 when his father asked him to consider running Bechtel. After a few weeks of soul-searching, he became convinced that he was ready for the job. And he wanted to do it solo. "If you want me to take over," he told his father, "I will. But I'll have to do it my way. When I take over, I'm the boss." Without hesitation, Steve Sr. agreed, and father and son shook hands on their new alliance.

As a manager, Steve Bechtel Jr. differed greatly from his father. While Steve Sr. virtually carried the company around in his head and relied on the institutional memory of trusted lieutenants, Steve Jr. was in every sense the first professional manager to run the company.

Steve Jr. had a grip on the numbers and the details of the business in a way his father never had. Said Ed Garbarini, president of Bechtel Power Corp., "Steve Sr. was imaginative, intuitive, instinctual. He was the best salesman who ever came down the pike, the best business development person the company ever had. . . . But he never had the patience with numbers that Steve Jr. has. . . . Steve Jr. has a terrific ability to analyze figures and interpret them into meaningful data that he can use to make decisions. He goes through numbers like a calculator."

In 1960, Bechtel needed this kind of focused leadership. The company's rapid growth and its expanded scope had overwhelmed its own internal controls systems. The new boss insisted that detailed records be kept and major decisions be documented so that the pattern of decision-making was clear. Steve Jr. enhanced the company's planning capability and helped the more hidebound managers to understand the necessity for change. At the same time, the country and the corporation were entering a new era. New forces—environmentalism, globalism, economic upheaval, and intensified international competition—were to mark his term as leader of Bechtel.

To say that Steve Jr. was disciplined is to say that Hoover Dam had a little concrete in it. Steve spent a third of his time traveling, always accompanied by background material—covering both business and personal history—on everybody he was going to meet. "The one thing he really dislikes," his longtime personal assistant, Hart Eastman, once said, "is being surprised."

Though the details were his guidelines, he never got lost in them. Steve Jr. was instrumental in applying team management principles and in the process changed a company that had been a top-down hierarchy. Perhaps because of his early contact with so many Bechtel veterans, he believed that through collaboration, by exposing all workers to those with superior talents, coworkers become inspired and motivated. "It permits common men to do uncommon things," he said.

This is somewhat ironic, because Steve Bechtel Jr. was by no means a common man. In less than two decades, he doubled the size of an organization that his forebears had taken 60 years to build and transformed Bechtel into a modern company.

He had the uncommon ability to shift easily from studying technical details to setting overall strategic goals, from dealing face-to-face with craftspeople to meeting heads of corporations and even nations, and from disciplined analysis to inspiring encouragement. Besides transforming Bechtel, he contributed greatly to his profession and to society through many activities. His legacy will reach beyond Bechtel for many years to come.

Steve Jr. was awarded the National Medal of Technology by President Bush, one of the many honors of his career.

CHAPTER FIVE

A DECADE OF MEGAPROJECTS

Extending Global Reach

1970–79

The rules began to change in the 1970s. These changes were driven by major shifts in global economies and the gradual unraveling of the post–World War II order. The first Earth Day, celebrated in April 1970, would affirm broad public sentiment in favor of more stringent environmental safeguards. In the United States, that sentiment was given regulatory teeth in July, when President Richard M. Nixon created the Environmental Protection Agency.

In the realm of global business, the nations that made up the Organization of Petroleum Exporting Countries (OPEC) flexed the cartel's muscle and raised oil prices by more than 100 percent in 1973. It was an act of instant inflation that shocked the economies of most of the world's developed nations. By decade's end, inflation in the United States had reached double-digit proportions. Bechtel executives found themselves managing in a drastically altered business milieu, where the parameters could change at any minute, where social and environmental considerations could take precedence over growth and profits.

The company did not have to acquire a sense of environmental responsibility. An avid hiker and outdoorsman, Steve Jr. had a personal concern for the environment. "When we did the Trans Mountain pipeline in the early 1950s," recalled Steve, "it was the first pipeline across the Rockies, and our environmental concern was major."

Steve Jr. realized, as did Bechtel managers, that the much tougher

1971

Bechtel begins construction management and detailed design of the Washington, D.C., Metro transit project, which continues into the late 1980s. Of the system's 98 miles, 38 will be in the District of Columbia, with 30 miles extending into the Maryland suburbs, and an additional 30 stretching out into surrounding Virginia communities.

OPPOSITE: *The James Bay hydroelectric complex was begun in 1972. This monumental project involved the construction of four power stations and the diversion of four rivers. It increased by 10 million kilowatts the power available to Quebec and remains the largest civil engineering project ever undertaken in Canada.*

The Smithsonian Metro station, like others in the District of Columbia, features the high coffered ceilings, air-conditioning, and recessed lighting characteristic of the Washington Metro system.

1972

Bechtel helps manage and coordinate work on the huge James Bay hydroelectric complex; work continues into the mid-1980s.

environmental regulations were not only a social mandate but also a business opportunity. "Our primary effort should be in-house," he said, "to build up the expertise and capacity to help clients meet environmental considerations in siting, designing, and building facilities." For instance, an environmental laboratory established in Belmont, California, laid the groundwork for later heavy involvement in hazardous-waste handling and treatment. The decontamination and decommissioning of Three Mile Island would be the most visible such assignment.

Succeeding in these times required a company to be socially aware, politically savvy, and totally flexible—in short, to be a kind of company different from what Bechtel had been. Remaking Bechtel into a nimbler organization dominated discussion of the company's first five-year plan at the April 1970 meeting of the company's Directors' Advisory Group. The plan was built on four major projections for the half decade ahead: Business would expand, the international workload would grow, revenue and profit increases would shift to nonenergy-related areas, and revenues from big jobs would increase.

The company's 1970 decision to expand its international business and to go after big jobs was prescient, since the U.S. economy would soon go through hard times.

TAMING THE NORTH SEA

When the price of crude rose to about $40 a barrel, finding new reserves of oil and gas became a most attractive economic prospect. After all, the fear was that the price was headed for $100. With Europe desperately seeking indigenous energy sources, the North Sea, with its known but untapped deposits of oil and gas, became the world's most promising area for new energy exploration. And with the signing of a contract to develop the Argyll field in 1972, Bechtel entered North Sea history.

When Hamilton Brothers contracted with Bechtel to develop Argyll in the British sector of the North Sea, it wasn't the biggest, most technically complex, or the most costly job Bechtel had ever undertaken. But Steve Jr. would always think of it as a landmark because Argyll affirmed Bechtel's role as a leader in petroleum production technology.

Bechtel's experience in oil platform work dated to the early 1950s with Saudi Arabia's Safaniya field, then the world's largest offshore development. But as one Bechtel marine specialist said, "Those platforms, situated in relatively shallow

1972

The Bechtel-built Ponce petrochemical complex, the largest chemical plant built to date, opens in Puerto Rico. It is capable of producing 3 billion pounds of chemicals each year.

1972

Bechtel begins work on the Argyll and Piper oil field platforms in the British North Sea, some of the roughest waters in the world. The projects recover oil from thousands of feet beneath the seabed.

Puerto Rico's Ponce petrochemical complex produces everything from antifreeze and plastic sheeting to synthetic fibers and solvents for paints.

water, were dwarfs compared to the 40,000-ton giants in the North Sea." Within three years of securing the Argyll job, Bechtel was at work on three large North Sea projects—Argyll and the Occidental Group's companion Piper and Claymore fields—all located in churning, frigid waters more than a hundred miles off Scotland's rugged northeast coast.

Never before had oil been recovered from fields thousands of feet beneath a seabed, which itself lay 480 feet below some of the world's roughest waters. The prize was a bonanza in light, easily processed crude.

In the North Sea projects, design and construction technology was constantly evolving, and each new platform seemed to incorporate a major breakthrough. Argyll was the first anchored semisubmersible drilling rig converted to service as a production platform. Piper was the largest offshore structure ever launched at sea from a barge. Claymore was linked to Piper by a 30-inch submarine pipeline that enabled them to work as a coordinated pair.

Constructing the 30-inch Piper pipeline—the largest ever at this depth— was a major achievement because bad weather limited the 1974 construction season to just 131 days. The 14,000-ton Piper platform was a worldwide effort. Pilings came from the United States, bottom leg bottles from Japan, platform superstructure from the Netherlands, much of the drilling equipment from Canada, the upper part of the jacket from France, and the bottom-half template and pumping and production modules from Scotland.

And there was a huge amount of additional work to come: the Beryl "B" platform for Mobil; a project services contract with Conoco for the Murchison platform, the most northerly field developed at the time, 120 miles northeast of the Shetland Islands; the Hutton platform, with its revolutionary tension legs that enabled it to operate in waters twice as deep as any previously worked; and the Flotta oil and gas terminal in the Orkney Islands, which began operation in 1976.

As immense and important as the British sector work was for Bechtel, many thought the potential for future North Sea offshore work was even greater in the Norwegian sector to the northeast. One was Israel Leviant, former Bechtel vice president, director, and veteran European consultant, who had lobbied long and hard to promote this belief within Bechtel, initially with little success. "The Norwegian offshore," he said, "is comparable to Saudi Arabia. If you take the length of the Norwegian coast and make it straight, it will go around the Earth.

In São Paulo, Latin America's largest city, Bechtel is hired to provide engineering and start-up services for a new metro system, the most technologically advanced in the Southern Hemisphere.

The mammoth Murchison platform in the North Sea was built to withstand 90-foot waves and 110-knot winds.

OPPOSITE: *The Piper field platform in the North Sea off the coast of Scotland was, at the time, the largest structure ever launched at sea from a barge.*

They are just beginning to scratch the surface of it, and there will be several decades of work there." That promise began to be fulfilled when massive tugs guided into place the unwieldy Gullfaks "A" oil and gas production platform, a facility benefiting from project planning, engineering supervision, and construction and hook-up assistance by a Bechtel joint venture. Tall as the Eiffel Tower, the gravity-base platform was towed upright some 150 miles offshore Norway. It began producing oil in 1987 as part of Norway's planned multibillion-dollar development of its North Sea petroleum resources.

SYNCRUDE: OIL FROM SAND

In the 1970s, as oil prices skyrocketed, Canada resolved to fully develop its abundant domestic energy reserves. Syncrude Canada Ltd. enlisted Canadian Bechtel Ltd.'s help to extract synthetic crude oil from the Athabasca tar sands—a desolate 21,000-square-mile area in northern Alberta, where Bechtel had helped develop the Suncor project in the early 1960s.

Much of the work on the new project was performed 300 miles to the south of Athabasca, in Edmonton. At Bechtel's field operations office there, skilled tradesmen drawn from every Canadian province preassembled modules and component parts on a scale never before attempted. Included in the 800,000 tons of equipment preassembled in Edmonton and trucked to the job site were more than a million linear feet of pipe. This massive preassembly was a key reason the project team completed the $1.8 billion Syncrude complex within budget and four months ahead of schedule.

When construction was completed in 1978, the Syncrude facility was one-third larger than any other mining operation in the world, it was the biggest plant of its kind in the world, and it included the largest fluid cokers ever erected. Most important for Canada, the facility produced up to 125,000 barrels of high-grade synthetic crude oil daily—nearly 10 percent of the country's total oil production.

His Majesty King Khalid and U.S. Treasury Secretary Michael Blumenthal inspected a model of Jubail at the dedication ceremonies in 1977.

JUBAIL: THE GIGAPROJECT

Never has there been anything like Saudi Arabia's Jubail Industrial City: a self-contained complex built from the ground up over a 20-year period to

accommodate as many as 370,000 people. The Jubail program would include all the requisite infrastructure, roads, sewage and irrigation systems, and schools.

Jubail owes its existence to the fact that most of Saudi Arabia's oil fields are charged with abundant quantities of natural gas under high pressure. Traditionally, the standard procedure was to burn off this gas. Bechtel legend has it that while flying into Dhahran from Beirut, as he did so many times, Steve Sr. would look down and see the flicker of bright orange-red flares dotting the desert below like dozens of Bedouin campfires and wonder about ways to utilize this energy resource. There was little economic incentive to move the gas to world markets. "It had been common knowledge," he said in a 1983 interview, "that the flaring of gas

Jubail, formerly a tiny fishing village in Saudi Arabia, is now a large city complete with schools, housing, roads, railroads, an airport, a telecommunications system, an industrial port and harbor, and a self-contained industrial complex.

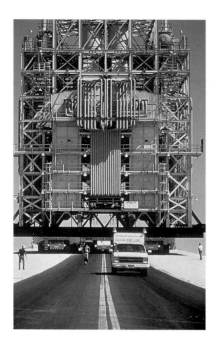

Prefabricated modules for an industrial facility were built off-site, shipped, and then moved by heavy-lift transporters to permanent sites in Jubail, the Bechtel-built city in the sand.

was going on all over Saudi Arabia. It was a question of what they should do with it, how best the Saudis could utilize this excess energy for the development of their country. My initial concept was for the design and construction of an industrial complex in the Eastern Province of Saudi Arabia, to include such facilities as refineries, petroleum and petrochemical plants, and steel, fertilizer, and aluminum production facilities." These were all heavy energy users that would devour the excess gas.

Planning for the industrial development began in earnest on April 27, 1973, when Steve Sr. called a meeting to outline his ideas. Among those present were two men who would play key roles in the ultimate success of the project: Suliman Olayan, Bechtel's longtime associate in Saudi ventures, and Parker T. Hart, a Bechtel consultant and former ambassador to Saudi Arabia.

The best location for the complex was Jubail. A fishing village of around 7,000 people, Jubail was the only undeveloped port in the Eastern Province with access to deep water. It lay close to developed areas such as Ras Tanura, Ad Dammam, Al Khobar, and Dhahran, and just southeast of the major Berri oil field. A deep-draft port could be built at Jubail to handle ore carriers for the proposed steel mill and bulk cargo carriers that would be needed during construction and operation.

Steve Bechtel Sr. presented his plans for the plant to King Faisal on the afternoon of June 4, 1973, in Geneva. At the meeting, the personal rapport between the two was instantly apparent. Steve had been a friend of Faisal's father, King Abdul Aziz, and had worked with him to develop his country. And Faisal himself had visited the Marinship facility during World War II, with Steve Sr. as his personal guide.

Along with new industry, a planned city would provide housing and residential amenities. A power plant of perhaps 1,000 megawatts, provided in staged increments, and a ready supply of cooling water would be critical elements of the infrastructure. A third key element would be gas-gathering facilities and pipelines to bring natural gas from the production fields to the industrial area.

The Saudis were determined to include, as part of the project, a strong manpower training program to prepare their people for all sorts of tasks. Dr. Jameel Al-Jishi, the first director general of Jubail, emphasized the training of Saudis. "If all we get is the buildings and the infrastructure and you leave, what will we have?" he asked. "But if you teach my people, when you leave, we will have a city. And this is what I want." Every department or project manager had a Saudi counterpart, and, at some point, the Bechtel employee would no longer be needed.

The Saudi government signed a contract on June 2, 1974, calling for Bechtel to oversee a long-range industrialization program on behalf of the kingdom. It was thought that the work could go on for 10 years or more. Jubail would be the principal site for industrial development in the kingdom, and it would be paired with a smaller development at Yanbu on Saudi Arabia's Red Sea coast.

The effort showcased the company's growing ability to manage complex projects. Even for a company where multibillion-dollar deals would become commonplace, the ultimate numbers on Jubail were awesome. "Before any industries or permanent housing could be built, we had to install all the infrastructure needed to support it," said Project Manager Mort Dorris. That meant raising the mean elevation of the site and moving more than 440 million cubic yards of earth, enough to build a road around the Earth at the equator nearly 30 feet wide and

In the Jubail industrial park, 16 primary industries operate around the clock, creating more than 30 different products. Among the facilities are a fertilizer plant, an oil refinery, a steel mill, and two methanol plants.

more than 3 feet deep. Power and water supplies had to be installed, with generators, wells, and small desalination plants gradually replaced by a national system. A total transportation system was built from scratch—a network of national highways, an airport, and a huge port complex.

Bechtel's initial concept was to serve as the agent for the Royal Commission for Jubail and Yanbu, vested with full authority for executing contracts and free to do detailed engineering-construction for specific projects. The final contract precluded Bechtel from acting as a vendor or contractor, but allowed it to perform detailed engineering when requested by the Commission, and gave it authority over contract management, administration, and a few other specified areas. On June 24, 1976, a 20-year program management services agreement for the development of the Jubail region was signed in Riyadh.

The first primary industry, the steel mill, came on-line in 1982. By the time a 15-year development review was completed in 1992, there were 16 primary industries producing steel, fertilizer, plastics, and a range of petrochemicals and petroleum products. This vast and constantly expanding industrial base, along with more than 70 secondary and support companies, has transformed Jubail from a sleepy fishing village into a major force in world petrochemical markets.

NEW PHASE: FROM CONSTRUCTION TO FOCUS ON MANAGEMENT

Bechtel had evolved into a full-service engineering and construction company. If the customer desired, Bechtel could develop the concept for a project, perform feasibility studies, help arrange financing, perhaps invest some of its own money, do the engineering, and either perform or manage construction. By the mid-1970s, engineering and construction would be increasingly divided. Often, Bechtel would do neither the engineering nor the construction, but would serve as project manager, representing the owners.

A good illustration of this shift was Bechtel's involvement in the massive James Bay complex in a remote and inhospitable region of Quebec. The largest hydroelectric project in North America, involving the diversion of four rivers into one and the construction of four huge powerhouses, it took 13 years to complete, beginning in 1972. Bechtel's main role was to help manage the activities of a workforce numbering some 12,000 people and to help oversee the supply of hundreds

1974

Planning begins on international airports in Saudi Arabia, including King Khalid International Airport and King Fahd International Airport. The King Khalid project includes five terminal buildings, a control tower, a mosque, a ceremonial mall, and support and utility buildings, as well as a self-contained community for 3,000 people.

OPPOSITE: *The massive King Fahd International Airport near Dammam is Saudi Arabia's third international airport, with a passenger terminal and a mosque designed in traditional Islamic style.*

Construction begins in Brazil on SAMARCO, the largest capacity long-distance iron-ore slurry pipeline in the world at 7 million metric tons a year. The pipeline transports iron ore concentrate from a mine north of Rio de Janeiro to shipping facilities on the Atlantic coast 250 miles away.

1976

Bechtel begins construction of the Palo Verde nuclear generating station near Phoenix, Arizona. Completed in 1987, the plant uses reclaimed sewage water for the cooling towers and is the largest electrical producer of any kind in the United States.

of thousands of tons of equipment and materials to a site more than 400 miles from the nearest town.

On New Year's Eve 1975, the company was at work on 119 major projects in some two dozen countries with estimated values, according to *Fortune* magazine, totaling $40 billion. For the preceding 20 years, it had been growing at the astonishing rate of 10 to 20 percent a year. And it would continue to expand at or above that pace through the 1970s.

Bechtel was clearly entering a new phase in its evolution, and major change seemed inevitable. Its success had generated cash reserves, which needed to be invested.

In 1977, the company became part of a consortium that bought Peabody Coal Co. from Kennecott Copper Corp., its first non–real estate equity investment. In 1979, a Bechtel/Hanna combine invested in Welltech, Inc., an oil-well-servicing business in Houston. That was soon followed by the acquisition of Dual Drilling Co. Investments in such companies as Mesa Petroleum and Lear Petroleum led to a position in oil and gas exploration and development. Bechtel formed Uranium Enrichment Associates (with Union Carbide and Westinghouse) and Energy Transportation Systems, Inc. (a coal slurry pipeline project with Lehman Brothers), development companies whose missions were never brought to fruition.

These new investments were all in industries that complemented Bechtel's construction core and would not put the company in a competitive position with any customers. Said Steve Jr., "We want to get into things where we have more to bring than just money, things that won't confuse our primary business."

As the decade progressed, Bechtel's outside activities were accompanied by the creation within the company of a series of major new enterprises, including Saudi Arabian Bechtel Co., Bechtel National, Inc., Thermal Power Organization, Bechtel Financing Services, Commercial Buildings & Land Operations, and the Nuclear Fuels Operation. Each was designed to capitalize on Bechtel's increasing specialization in a variety of fields. Characteristic of this carefully targeted effort to address new markets and shifting demands were the creation of Bechtel Energy Corp. and the acquisition and expansion of Becon Construction Co. Over the following decades, Bechtel would continue to adapt to constantly evolving business environments with new collaborations and services.

Algeria's Arzew LNG facility can liquefy more than a billion cubic feet of natural gas per day. Once liquefied, the gas takes up one-six-hundredth of its original volume and can be exported economically aboard specially built tankers.

1976

Bechtel signs an agreement with Saudi Arabia to develop the Jubail project, a 20-year plan to create one of the largest civil developments in history. Located on the east coast of the Arabian Peninsula, the project includes the development of industries (petrochemical, fertilizer, and metal processing), a major harbor and port facility, a national airport, public service utilities, roads and highways, rail lines, telecommunications systems, and a city that ultimately is expected to house 370,000 residents.

1976

Bechtel joins with Lehman Brothers and others to form Energy Transportation Systems, Inc. and develop a proposal for a coal slurry system running more than 1,000 miles between Wyoming and Arkansas.

A NEW GENERATION OF LEADERS

Given all these changes, it was logical that Bechtel of the 1970s would be led by a new generation of managers. Between 1967 and 1977, 12 of 15 board members retired. Among the new people brought in were generalists who had no experience in the construction business but had broad and impressive backgrounds in public administration, the academic world, and private enterprise.

Chief among that group was four-time Cabinet officer George Shultz, whom Steve Jr. first met while giving a speech in Washington. Shultz had been impressed enough to call Steve from time to time for advice or simply to sound him out on a particular problem. In May 1974, Shultz left government service—he was by then secretary of the treasury—to become executive vice president of Bechtel Corp. A year later, he would become president.

1978

Algeria's first LNG plant is dedicated. Built by Bechtel, the plant triples Algeria's export capacity of liquefied natural gas. Gas produced in Algeria is liquefied and loaded aboard nine specially built tankers for export.

1979

Bechtel assists in the cleanup and recovery of the Three Mile Island nuclear power plant, developing several generations of robotic devices for use in decontaminating the damaged unit.

In addition to Shultz, board members joining Jerry Komes, John O'Connell, and Harry Reinsch on Bechtel's executive committee in the mid-1970s included Porter Thompson, Steve White, and George Saul. More than ever since it had been established in 1959, the committee served as a high-level forum for review, discussion, and formal decision-making.

Bechtel would need the considerable talents of all its leaders in the next decade. The year 1979 started badly and then got worse. The lack of a U.S. energy plan, a slowdown in the demand for electricity, and increased financial restraints led to an epidemic of project cancellations and delays. Then, on March 28, an accident occurred in a nuclear plant at Three Mile Island, near Harrisburg, Pennsylvania, which terrified the nation and contributed to the demise of the market for nuclear power plants in the United States. Later, Bechtel would take on the cleanup of Three Mile Island—a facility in which the company had no prior involvement.

Bechtel was, by tradition and competence, a direct construction company, yet nonconstruction activities such as project management, engineering, and construction management now accounted for two-thirds of all revenues, up from 40 percent in 1970. The triumvirate of companies that bore the Bechtel name still ranked No. 1 among U.S. engineering and construction firms, but a very difficult decade lay ahead.

OPPOSITE: *Solid ice and solid steel coexist in Greenland's Qaumarujuk Fjord. A mile-long aerial tramway links the Black Angel mine with the concentrator on the opposite shore.*

CHAPTER SIX

GLOBAL DOWNTURN

Survival

1980–89

The new decade got off to an auspicious start as Bechtel posted $11.3 billion in new bookings for 1980, more than double the previous year and the highest ever, to that point. Crews were at work on 132 major projects in 20 countries. Anticipating further growth and expansion, the company underwent a major corporate restructuring at the end of 1980. With new work at an all-time high and a new management team in place, Bechtel seemed well-positioned for the future. But tough times lay ahead.

In 1981, the economy slid into the sharpest and deepest recession of the post–World War II era. A December 1981 cover story about Bechtel in *Forbes* magazine laid out an ominous scenario: "A sharp worldwide recession can hurt an engineering company with its high overhead in salaries. . . . The potential for losses, as competition heats up internationally and customers demand fixed-price bids, is haunting. Losses can quickly erode the thin capital bases found in this, or any other, service industry. . . . There is daily risk, hourly risk all across the globe."

Many business leaders, Steve Bechtel Jr. among them, worried that the economic difficulties of the early 1980s were exacerbated by a rising tide of overregulation that increased project costs at every turn. For the construction industry, the impact of this potent mix of political forces in the midst of a worldwide economic downturn was devastating.

OPPOSITE: *Bechtel engineers were brought in to Three Mile Island after the accident and worked with extreme caution to clean up the site. The basement of the containment building that housed the damaged reactor was filled with a mass of radioactive water and rubble. It was 16 months before anyone could go into the reactor vessel.*

Bechtel technicians took soil samples in a Glen Ridge, New Jersey, neighborhood. Bechtel designed the cleanup strategy for several communities affected by radioactive waste from a defunct watch dial factory.

1981

Bechtel is awarded the first of many contracts for the Formerly Utilized Sites Remedial Action Program (FUSRAP), to reduce low-level radioactivity and treat hazardous waste remaining at U.S. sites primarily associated with the Manhattan Project.

1982

In New Zealand, Bechtel begins construction of the world's first commercial facility to convert natural gas to gasoline.

SHIFTS IN LEADERSHIP AND MANAGEMENT

To meet the new challenges, a new holding company, Bechtel Group, Inc., was formed. As the twin pillars of this organization, Steve Bechtel Jr. assumed the title of chairman and CEO and George Shultz was named president. Shultz made his own major contributions to changing the way Bechtel viewed its business. Since his arrival, he had been educating the senior managers at Bechtel, convincing a tight-knit group of project-oriented executives that they had to work as hard at managing their cash flow and financial investments as they did at managing their construction projects. He helped reorganize Bechtel's investment schemes, including Sequoia Ventures, which was active in real estate and energy projects. Little acorns, Shultz called them. "We can afford," he said, "as a private company, to make investments we have to be patient with."

Shultz returned to government as secretary of state in 1982, setting off a fresh round of media speculation about Bechtel's influence in Washington. Steve reflected on the attention brought to Bechtel: "It bothered me for awhile, particularly the insinuations about our supposed influence with the government. It's unfortunate that we don't have all the power it's alleged we have. If we did, we could help fix some of the problems that exist around the country." The following May, after Shultz's departure, Steve announced that Alden Yates would succeed him as president of Bechtel Group, Inc. and a member of Bechtel's executive committee. "A partner in the company ever since he got out of school," is how Steve introduced Yates, whose father, Perry, was a close associate of Steve Sr. and an original executive committee member.

GLOBAL CHALLENGES

The economy continued to wreak havoc on business regardless of who was in charge. BART and the Washington Metro jobs had led to major construction management contracts in Atlanta, São Paulo, and Caracas, but no more big transit projects followed. Hotel construction tailed off, and the power market, which had been Bechtel's biggest moneymaker, collapsed. New utility awards fell from 233 gigawatts during the 10 years from 1965 through 1974 to 55 gigawatts in the 13 years from 1975 through 1987. Not a single new U.S. nuclear power plant had been ordered since 1973. Bechtel's power workload, which had been increasingly dominated by nuclear, plummeted as Three Mile Island, and later Chornobyl, heightened public concern about nuclear power, and increasing government regulations sent capital costs through the roof. At the same time, North Sea oil, the Alaska pipeline, and price-sensitive conservation measures combined to reverse the rise in oil prices and doom a variety of alternative energies and synthetic technologies. Finally, a host of less-developed countries had become so mired in debt that servicing existing loans became problematic and taking on additional debt for new projects was out of the question. Future growth for Bechtel would come from smaller, more diverse projects such as cogeneration, toxic cleanup, and power plant maintenance.

As the U.S. recession spread to the global economy, Bechtel faced tougher competition for overseas work from non–U.S. firms that were often supported by their own governments. The industry suddenly had a number of

A module for New Zealand's gas-to-gasoline plant was built in Japan, then shipped and trucked to the job site. Bechtel assembled an international network of suppliers to construct the plant in record time for a New Zealand project.

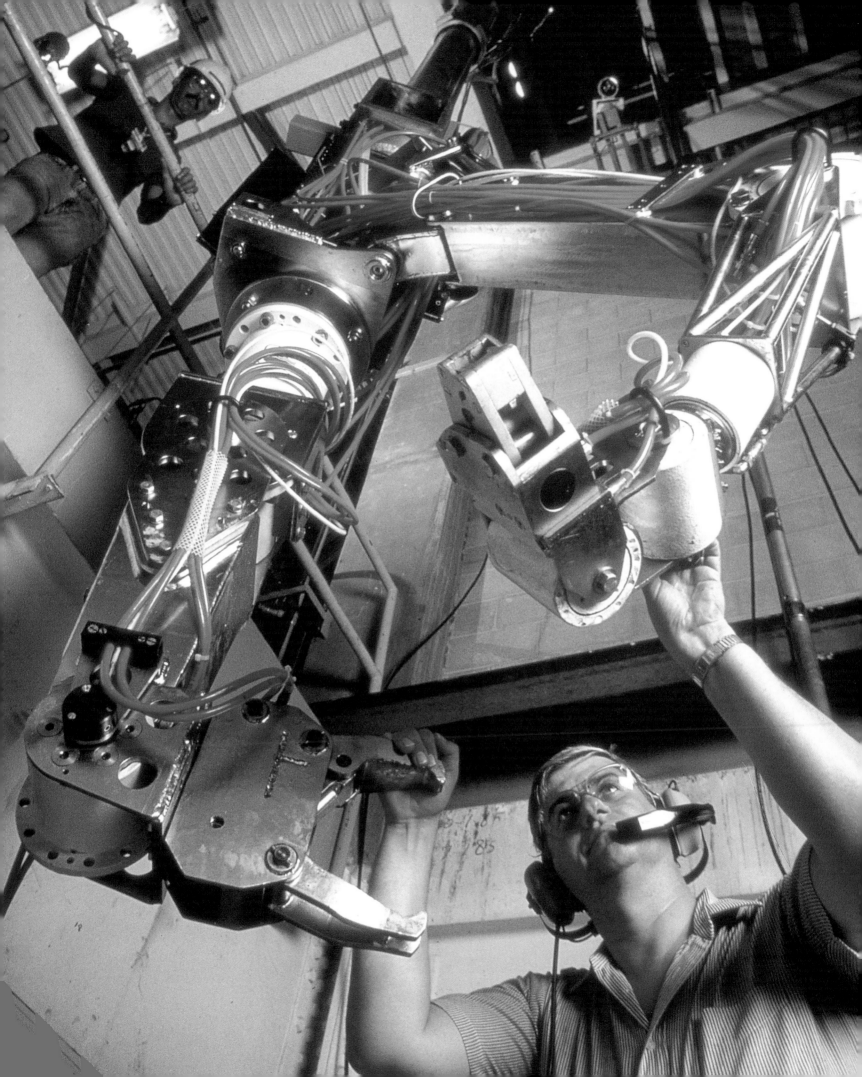

new players, new companies, new countries, and new governments that wanted to take on work themselves, seeking outside help to train local workers and to transfer technology.

Bechtel became increasingly multinational. Senior managers, each of whom had spent years traveling and working around the globe, understood that they were outsiders and had to make some attempt to help develop the countries where they worked. The company increased the number of foreign nationals in the organization and opened regional offices to develop local engineering competency. "We have to look at ourselves as a truly international organization," said Steve Jr. "We have to approach projects as a multinational organization with a multinational staff and multinational sources."

THREE MILE ISLAND CLEANUP

With few megaprojects in the offing, Bechtel established bailout teams—flying squads of engineers, scientists, and other specialists who could be rapidly deployed to emergency situations to assess what needed to be done and to do it. The teams were an outgrowth of the company's decontamination and cleanup business, which expanded after the company had taken on the billion-dollar cleanup of the nuclear accident at Three Mile Island.

In March 1979, the Unit 2 reactor core of the nuclear power station at Three Mile Island (TMI) experienced a partial meltdown that damaged the fuel rods and possibly other internal reactor components. Radioactive materials were released from the core into the reactor coolant system and into the reactor building. An avalanche of frightening questions came crashing down in response. What were the true dangers of the accident? How close could one get to the reactor without being exposed? What were the conditions in the reactor?

The first studies of the reactor revealed that radioactive isotopes had been flushed from the reactor vessel into the basement, and the levels of radioactivity in the 600,000 gallons of floodwater were higher than in the water of the reactor vessel. Before work could begin in the containment building, the atmosphere had to be cleared, the water cleaned out, and surface decontamination performed. In the summer of 1980, krypton gas was safely removed. Despite an extensive decontamination effort, it was 16 months before a person could venture into the

1982

Bechtel oversees expansion of Los Angeles International Airport, the world's third-busiest passenger airport. The project includes two new terminals and a five-lane elevated roadway. In less than three years, the project doubles passenger terminal space and curb space for loading and unloading, just in time for the 1984 Summer Olympics.

1983

Bechtel provides start-up engineering and construction support for the completion and commercial operation of the Diablo Canyon nuclear power facility in California.

OPPOSITE: *Bechtel helped develop remote-controlled devices such as this robotic arm to collect information on radiation and carry out decontamination tasks in highly toxic environments such as the Three Mile Island Unit 2 containment building.*

Steam and electricity are generated from burning a coal-mining waste product called culm, once considered useless, at the Gilberton cogeneration plant in 1984. Bechtel's involvement in the plant reflected what environmental leaders later came to call "sustainable development."

OPPOSITE: *Because of its greatly reduced volume, liquefied natural gas can be shipped from production facilities to markets thousands of miles away, increasing its economic viability.*

building. By 1987, engineers from Bechtel and TMI owner General Public Utilities had succeeded in decontaminating the site, lowering radioactivity levels to establish a stable and secure facility.

OTHER PROJECTS

Despite an overall decline in its workload, Bechtel was still able to win important projects in the U.S. market and around the world. Bechtel Energy Corp., specializing in public power projects, continued to make appreciable gains, as did Becon Construction in open-shop construction. New power contracts included:

 – A coal-fired plant for Gulf States Utilities near Lake Charles, Louisiana.

 – A massive oil-fired generating complex for CADAFE, a government-owned utility company, on Venezuela's Caribbean coast.

 – A coal gasification combined-cycle plant for Southern California Edison and Texaco near Barstow, California.

Increasingly, new work resulted from Bechtel's technological prowess:

 – In New Zealand, the world's first commercial facility for converting natural gas to gasoline was built.

 – A refuse-to-energy plant at McKay Bay near Tampa, Florida, was the first to employ Danish combustion technology for converting waste to energy.

 – A Rochester, Massachusetts, waste-to-energy plant incorporated innovations in cooling-tower technology to reduce water consumption.

 – A California plant was the first to use coal-derived gas to drive gas and steam turbines in tandem.

 – The first production-scale laser-isotope uranium enrichment facility was developed by Bechtel.

PRUDHOE BAY

Two of the more demanding projects of the period were the design and construction of an offshore seawater treatment plant for Arco (Atlantic Richfield Co.) and a low-pressure separation facility for Sohio (Standard Oil Co. of

Liaison International © Lincoln Potter

1984

The Summer Olympics in
Los Angeles begin. Bechtel
executives help plan the
conversion of dozens of
sports and training locations
as well as housing facilities
at three college campuses.

1984

The first gold is mined at
the Ok Tedi gold and copper
project. Bechtel, in a joint
venture, provides engineering, procurement, and
construction services. The
project mines, processes, and
ships more than a million
tons of copper ore from an
isolated mountaintop in
Papua New Guinea between
1987 and 1994 alone. The
project also includes mining
high-grade gold that lies
directly on top of the copper.

Ohio). Both projects would boost crude oil recovery at Prudhoe Bay on Alaska's North Slope, a resource-rich area containing an estimated 9.4 billion barrels of oil and 26 trillion cubic feet of natural gas.

The seawater treatment plant was designed by Bechtel and built at a ship-yard in Korea before being towed across the Pacific and Arctic oceans in mid-1983. It was the largest piece of equipment ever transported to Prudhoe Bay. The 26,000-ton, 610-foot-long plant was designed to treat 2 million barrels of seawater daily, which would be injected into Arco's oil reservoir to maintain its pressure and enhance production levels.

The $600 million low-pressure separation plant consisted of a series of pre-built modules, some the size of 10-story apartment buildings. The modules were built in Richmond, California, loaded onto barges, and shipped to Alaska. Once installed and operational in 1984, the facility separated gas and water from crude oil recovered from Sohio's Prudhoe Bay oil field.

Because the harsh weather conditions at Prudhoe Bay made on-site assem-bly prohibitively expensive, each module had to be a fully constructed, enclosed building before being shipped to Alaska. Timing of the shipments was crucial, as the waters around Prudhoe Bay are navigable for only a few weeks each summer; they are completely frozen over the rest of the year.

OK TEDI

In 1981, Bechtel still found other mountains to climb, or mine—as was the case with the $1.5 billion Ok Tedi gold and copper project in Papua New Guinea. Here was a massive logistical problem, but the taming of hostile environments had been a part of the company's work since its earliest days. The problem in this case was getting supplies and equipment into the isolated region. So remote and forbidding was its geography that the local people had had no contact with the outside world until 1963.

The job for Bechtel was to build a gold processing plant and mine the gold and copper ore atop (and beneath) Mt. Fubilan, which was more than a mile above sea level. Despite wretched conditions and unprecedented drought, which evaporated Bechtel's 500-mile river-based logistics route, gold and copper ore were harvested in May 1984.

REMAINING RESPONSIVE TO A CHANGING GLOBAL ECONOMY

While interest rates soared, plummeting energy and commodity prices continued to reduce demand for major oil, gas, and mineral projects. Bechtel's top management responded on a number of fronts. In one important move, Steve created a new Group Strategy Committee, a rapid-response cluster that included himself and the heads of the three principal operating committees, to identify and react quickly to shifts in the changing construction market.

Throughout the company, developing new markets and carving new niches in old markets became major concerns. With utility construction stalled, Bechtel Power vigorously pursued operations, maintenance, cogeneration, and geothermal work. With nuclear power construction withering on the vine, Bechtel helped troubled power plants find third-party financing to complete construction. Bechtel Civil & Minerals upgraded its nuclear fuel operation into an Advanced Technology Division that could compete for toxic waste management projects, decontamination work, and advanced engineering for military and space programs. That

1985

In one year, Bechtel builds a 300-mile, 24-inch pipeline in Colombia that will move crude oil from the Caño Limón field in the interior, across the Andes, through rain forests, under rivers, and across plains and swampy terrain to the Caribbean port of Coveñas.

organization would soon be folded into Bechtel National, Inc., led by Bill Friend. Overseas, Bechtel had joined with a Taiwanese firm to establish Pacific Engineers and Constructors, Ltd. and formed Bechtel China with headquarters in Beijing.

Underscoring the value of technical excellence, Bechtel established a Fellows program in 1984. Bechtel Fellows advise the company's senior management on policy issues and report to them on key developments in their fields—from geophysics and hydraulic engineering to computer animation and environmental remediation.

A major effort was launched to fine-tune competitiveness, and prices were pared to maintain market share. This was a different mindset. "From the mid-1960s until the early 1980s, we had an unprecedented period of prosperity for the construction industry throughout the world," noted Executive Vice President Harvey Brush. "There was a lot of money and very high demand for extractive industry, manufacturing, and infrastructure." The customer's major incentive was to get the job done as quickly as possible and a kind of laissez-faire attitude had developed toward total project cost. "Schedule became all-important," added Brush. "It dictated how you structured the job. Customers were concerned about technology transfer and training nationals, so that became part of the cost picture. People weren't paying as much attention to overall costs because they had other priorities."

Having developed much of its capability and senior management doing lump-sum work prior to 1967, Bechtel spent the following 15 years in a cost-plus world. Now, in the mid-1980s, construction was entering a new fixed-price era as markets tightened, and Bechtel found it could compete better by controlling the costs of labor and material. Howard Wahl, Bechtel Eastern Power Corporation's president, converted two major jobs from construction-management to direct-hire construction because he could better manage the cash flow and keep Detroit Edison within its budget. "It was a real learning process for both of us," said Wahl. "We had to learn to manage the project to the cash available and adjust the schedule instead of making the schedule and adjusting the budget accordingly."

KEEPING THE WHEELS TURNING

There were other lessons to be learned as well. Through much of the 1960s and 1970s, business had come to Bechtel not because of major sales development efforts, but because Bechtel had available a very large, very effective, and very efficient

The Portland, Oregon, Light Rail Transit project was designed by Bechtel to serve as a cost-effective way to ease traffic congestion in the rapidly growing city. Now greatly expanded, the original 15-mile system had 25 stops through downtown and the eastern suburbs.

OPPOSITE: *The Trans-Turkish Motorway is one of the largest highway construction jobs ever undertaken. As part of a system linking Europe with Asia, the highway across Turkey will transform the nation's economy.*

1986

Bechtel gets involved in completing the Channel Tunnel, connecting England and France. One of the century's largest engineering and construction projects and the world's largest privately financed infrastructure project, the tunnel opens in 1994.

1986

Bechtel begins work on a 70-mile portion of the Trans-Turkish Motorway, connecting Ankara and Gerede. A second 70-mile section, a beltway around Ankara, is scheduled to be completed in 1998.

system to get work done. "People looked at us as doers," said Harvey Brush. "That reputation was deserved, and we capitalized on it."

In the 1980s, division managers and business development people alike would discover that markets were adjusting. There would be more infrastructure work in the Third World. But increasingly, in places like Algeria and Indonesia, the emphasis was shifting from major industrial undertakings, such as LNG, oil, and gas, to human needs, such as agriculture, water, and housing. In Saudi Arabia, a new five-year plan (1986–90) marked a significant departure from the three that had preceded it, shifting emphasis from civil engineering to social development, from spectacular physical accomplishments to equally important human needs. The transformation reflected not only Saudi satisfaction with what had been accomplished thus far, but their realization that lower oil prices were going to be a fact of life for the foreseeable future.

Given the new constraints in the market, Bechtel clearly had to be more aggressive in its marketing. Once again, past was prelude. Steve Sr. had so many times cast his eye around the globe to see what was needed, and then shaped the job accordingly. In the 1980s, this practice was renewed with vigor. Said one executive: "If we don't have a client, we find one. If there's no project, we assemble one. If there's no money, we get some." But even that approach could not protect the company from the declining world markets. Bechtel, like many other companies, would learn a terrible new word: downsizing.

THE TRANS-TURKISH MOTORWAY

In 1986, the Turkish government commissioned Bechtel and its local partner, Enka, to build a section of a high-speed toll road across the Republic of Turkey that would be part of a transcontinental highway system linking Europe with Asia. Turkish highways were so jammed with cars and oxcarts that produce was perishing in transit, a loss that could be ill afforded by an economy half built on agriculture. The new roadway would contribute immeasurably to Turkey's economic development, permitting the transport of goods to Middle Eastern markets and to its own people. In October 1992, Bechtel and Enka completed the first phase of their work, a 70-mile, six-lane road connecting Gerede with Ankara. The second phase, a 70-mile ring road around Ankara, will be completed in 1998.

On May 27, 1986, Steve Bechtel Jr. announced an organizationwide realignment. Cordell Hull and John Neerhout Jr. were named executive vice presidents of Bechtel Group, Inc., seven new operating companies were created, and the London division was made a separate subsidiary called Bechtel Limited. The technical manager of the London operation became managing director of the new subsidiary. His name: Riley P. Bechtel. Noted Bechtel's terse announcement on the Business Wire, "He is a son of Chairman Stephen Bechtel Jr."

Over the next two years, Bechtel would continue to adjust to the new, more competitive world it was living in. "There's less work available," Steve said in 1988. "There are fewer megaprojects, more small projects, and tougher competition. And we anticipate the tough competition is going to last for several years, maybe longer." Even so, there were still the bell-ringer projects:

– In 1983, Bechtel was called in to complete the venues for the Los Angeles Summer Olympic Games. It was a last-minute plea from the Olympic organizing committee for Bechtel to help pull off the impossible and bring everything together after the process had begun to lag dangerously. At one site, workers had only 33 days to renovate an entire stadium.

– Between April 1985 and March of the following year, Bechtel constructed the Colombia pipeline on an extraordinary schedule. The 300-mile, 24-inch pipeline stretched across the Andes, including some of the most rugged terrain in South America. Engineers also had to construct an offshore loading terminal on the Caribbean while stationed on a converted supertanker. When completed, the pipeline provided a much-needed boon to Colombia's economy, tapping approximately 1 billion barrels of crude—enough to transform the nation from an oil importer to a major oil exporter.

RILEY'S BECHTEL

The business climate was still changing dramatically. And in 1989, that tumultuous external change would be reflected in a pair of significant internal events that would once again affirm Bechtel's ability to adapt to change through continuity. The year was not even three months old when, on March 14, Stephen D.

Bechtel Sr. died at the age of 88. A builder for more than 70 years, he had shaped the worldwide enterprise that his son now managed. Two weeks after the death of its senior director, Bechtel named a new president and chief operating officer. At the age of 37, Riley Bechtel became the fourth generation of his family to serve as president of the 90-year-old international engineering and construction firm. "Riley has grown up in our organization on job sites around the world, and comes to this position having served throughout the company for a number of years," said Chairman of the Board and Chief Executive Officer Stephen Bechtel Jr. "I know he has the spirit, the vision, the will, and the capacity to lead our operations into the next decade." With that endorsement, Riley Bechtel took charge of the family firm in his mid-30s, as had his father and grandfather before him.

1989
Bechtel Enterprises, the company's finance, development, and ownership affiliate, forms a partnership with Pacific Gas and Electric to create U.S. Generating Co. (USGen) to develop, own, and operate electric power facilities in North America.

1991
In a joint venture with Parsons Brinckerhoff, Bechtel begins construction on the Boston Central Artery/Tunnel project, America's largest infrastructure job. The plan includes a freeway beneath Boston and a tunnel beneath Boston Harbor.

OPPOSITE: *Bechtel designed, built, and, for years, operated the award-winning SEMASS waste-to-energy plant in Rochester, Massachusetts. When completed in 1989, SEMASS consumed nearly 2,000 tons of municipal solid waste per day, and an early 1990s expansion boosted capacity by half. Engineers used innovative technology to minimize the facility's water consumption and to achieve zero discharge of wastewater.*

RILEY P. BECHTEL

RILEY BECHTEL STILL HAS HIS first paycheck ($18.49 for a week's labor). Back in the summer of 1966, his father, Steve Jr., had offered to "cash" it for the 14-year-old, then had it framed and gave it to Riley as a memento. That first Bechtel job was as a mechanic's helper, or, as he says, "all-around gofer," sweeping floors, cleaning planes (including toilets), disassembling engines, and other odd jobs. "It was a great summer," he recalls.

He subsequently worked field jobs for Bechtel and others as a surveyor's helper, oiler, laborer, front-end loader operator, and carpenter. He also worked as a paralegal in Bechtel's Legal Department. "They were all great jobs," he said, "and I gained more *real* knowledge working than in most of my college and graduate school courses combined."

At the University of California at Davis, Riley majored in political science and psychology. After graduation and another stint in Bechtel's Legal Department, he continued on to Stanford, earning both a JD and an MBA. After interning at law firms and then passing the 1979 California bar exam, he was hired as an associate attorney for the San Francisco firm of Thelen, Marrin, Johnson & Bridges, Bechtel's longtime outside counsel.

Three of the four generations of Bechtel builders.
Left to right: Steve Sr., Riley, and Steve Jr.

In October 1981, after concluding that while law was both challenging and fun, his real satisfaction came from helping build things, Riley joined Bechtel full time. He was well aware (as were his bosses) of the gaps in his knowledge and was determined to learn from the bottom up. The first step in his education was as contracts coordinator for an Ann Arbor Power Division project; the second was as a piping superintendent.

In February 1983 Riley was assigned to Bontang, Indonesia, as area superintendent for a liquefied natural gas plant. Riley, his wife, Susie, and their two young sons arrived at the remote site to experience, on their first day, a strike by the 6,000 Indonesian craftsmen, and a big snake coiled around a light pole midway between their house trailer and the commissary. Riley says, "We had an absolute ball and learned a lot, including much of the language (in Susie's case) and the mandatory eight or so work-related phrases (in my case) a superintendent needs to know." His keen interest in the work, willingness to learn, and determined approach quickly won the respect of crew, colleagues, and bosses, who anointed him—with typical construction irreverence—CP, short for Crown Prince. Riley's Bontang stint would be fixed in his memory by a tragic accident, an explosion that killed three people and injured 50. He was several hundred yards from the explosion (not caused by Bechtel), which created an inferno and sent debris raining down. "I learned a lot about real leadership from my colleagues," says Riley, "as they worked to find and deal with the casualties, extinguish the fire, stabilize the plant, repair the damage, and complete the work with a jittery workforce."

In September 1983 (having helped get his "train" of the two-train LNG project from 15 percent to 90 percent complete), Riley and family moved to New Zealand, where he was general field superinten-dent ("No. 3 dog on the site," he says), for the gas-to-gasoline project. He says (again), "It was a superb experience and a lot of fun."

Riley and family moved to London in January 1985, where he served as a business development representative on prospects in Northern Ireland, Algeria, the Middle East, and South Africa. He then moved into operations and in early 1987 became managing director.

In August 1987, Riley was elected to the Bechtel Group, Inc., Board of Directors. Three months later, he was elected executive vice president and a member of the executive committee with responsibilities for all corporate functions and services and coordination of Bechtel's worldwide marketing and business development.

The company was touched by sad news in March 1989 when Steve Bechtel Sr. passed away at age 88, ending his special presence in the industry he had personally dominated for three decades. Two weeks later, grandson

Riley's hands-on experience in the field provided him with invaluable grounding for his later work.

Riley, then 37, was elected president and chief operating officer of Bechtel Group, Inc. Steve Jr. remained chairman and chief executive officer, and Alden Yates became vice chairman, a post he served in until he passed away later that year.

In June 1990, with his father's retirement at age 65, Riley was named CEO. Two months later, Saddam Hussein invaded Kuwait, trapping more than a hundred Bechtel employees and dependents in Baghdad. Through good fortune, effective crisis management, and the determination and commitment of all involved, the Bechtel employees and their families were released. Bechtel was able to continue its work in the rest of the Gulf region throughout the turbulence.

When Riley became chief executive, Steve Jr. became chairman emeritus, continuing the tradition he and Steve Sr. had established: the predecessor being supportively involved—staying informed and helping out, but taking great care not to interfere. Passing the baton to a fourth generation of family leadership is a rare, if not unique, occurrence in a premier, global corporation.

At its centennial, members of the fifth Bechtel generation are at work in the company learning the basics, earning their pay, and building their following. Bechtel is, of course, also blessed with a great depth of highly committed, capable, and talented nonfamily managers. Able leadership will continue to emerge over the next 100 years as it has during the last. Bechtel's future is bright.

CHAPTER SEVEN

BUILDING THE NEXT CENTURY

Leading Change

1990–98

Despite a worldwide construction slump that began in 1985 and lingered through the rest of the decade, Bechtel had managed to stay in the black, albeit at the bruising cost of deep companywide cutbacks. Bechtel competed hard and with some success. New work booked in 1989 climbed to $5.4 billion, a six-year high and a 21 percent increase over 1988.

BUSINESS ENVIRONMENT

Important market changes were under way. Globally, privatization was releasing telephone, utility, and other nationalized companies from state control. Europe was beginning to reduce economic barriers and to integrate its economies. With the collapse of communism, there was the potential for vast new infrastructure projects in Eastern Europe and the republics of the former Soviet Union. In Asia, the People's Republic of China was attracting foreign manufacturing and investment in infrastructure and tourism, and the "tigers" of the Association of Southeast Asian Nations (ASEAN) were racking up impressive economic growth.

There were, however, vexing challenges. Generally, the economically more developed nations, particularly the United States, continued experiencing the low-growth, low-inflation stagnation of the late 1980s, constraining investment. The indirect costs of increased regulation, again especially in the United

1990

The Hong Kong Airport Core Programme, which Bechtel helps manage, begins. The 10 projects include construction of bridges, tunnels, railroads, and an international airport at Chek Lap Kok, as well as extensive land reclamation.

OPPOSITE: *The mobile service tower housed the* Mars Observer *probe and its Titan III launch vehicle at Launch Complex 40 (LC-40), Cape Canaveral. It was the first rocket launched from LC-40, on September 25, 1992. Note the Bechtel logo on the Titan III's base.*

1991

Bechtel is tapped to help
complete facilities for the
1992 Summer Olympics in
Barcelona.

1991

Bechtel provides engineering
and management services
for the construction and
operation of the Athens
Metro project. The
excavation for two new
subway lines includes an
enormous archaeological
program.

*The overhaul and expansion
of McCarran International
Airport in Las Vegas will serve
the estimated 20 million
passengers who will travel
through the airport annually
by the year 2000. Bechtel has
worked at McCarran
throughout the 1990s.*

States, meant that otherwise viable projects were sometimes deferred or even canceled. At the same time, prices for natural resources such as coal and oil were decreasing for the first time since the Great Depression.

Quality-of-life issues for employees asked to move around the world, especially to remote job sites, had become significant. Bechtel, like most others in the industry, moved talented people to the work more frequently as projects were increasingly smaller and took less time to complete. Some of these employees moved with their families. Most went without them on "single-status assignments." Information technology improvements had begun to make possible moving more, if not most, design work to the people (instead of the converse)— provided distributed work processes could be conceived, designed, and implemented. Such basic changes in Bechtel's approach to its work would not be easy to achieve, could only start with some serious corporate "visioning and missioning," and would require significant investment without discretely measurable near-term returns.

It was clear that the 1990s would be a decade of intense competition and change, some rapid and all persistent. Floating exchange rates, integrated international capital markets, and improved communications were weaving Bechtel into a global economic web in which competition was becoming acute. Meanwhile, the broader industry uptick in spending for plant and equipment and government outlays for infrastructure that construction analysts had been projecting turned out to be slow in coming.

THE BECHTEL RESPONSE

Faced with these challenges, Riley Bechtel and his senior team resolved to meet the demands of the global economy and the listless business climate by managed change. With Bechtel resources increasingly deployed across the globe, the team focused its first efforts on sustaining and promoting "One Bechtel"—a spirit of teamwork and collaboration across business sectors and regional markets articulated by Steve Bechtel Jr. in April 1989.

With that foundation, they began a series of initiatives designed to improve the company's delivered value: continuous improvement and changes in work processes; refining the company's mission and strategies; restructuring the organiza-

CONTINUOUS
IMPROVEMENT

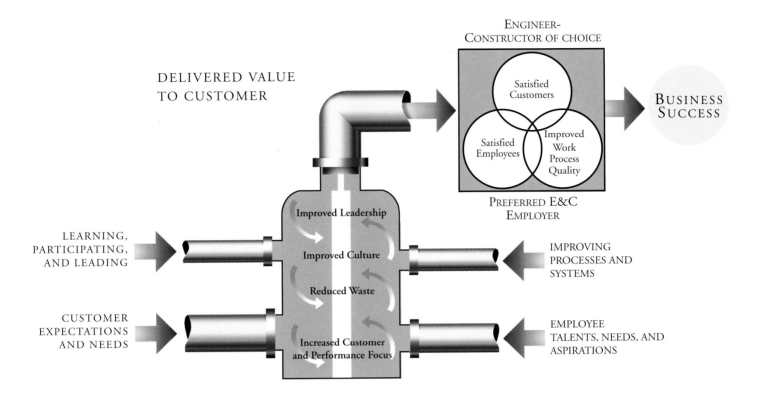

tional operating matrix to meet the needs of a complex business in a global economy; and improving leadership, strategic thinking, and market analysis.

Riley was keenly aware that a key component of Bechtel's success had been articulated decades earlier by Steve Bechtel Sr., who ascribed success to "real continuity of philosophy and objectives, strong traditions, and a consuming dedication to continuous improvement."

Bechtel sees continuous improvement as a closed-cycle process that yields delivered value. During the dramatic industrywide downturn of the 1980s, Steve Bechtel Jr. placed significant emphasis on reexamining and sharpening Bechtel's continuous improvement methodology. Riley and his 1990s management team would take the same approach. They depicted a refined approach to continuous improvement, as shown above.

But while Bechtel was making important process changes, world events were following their own course. In early 1991, the Gulf War tested Bechtel under the most challenging conditions imaginable.

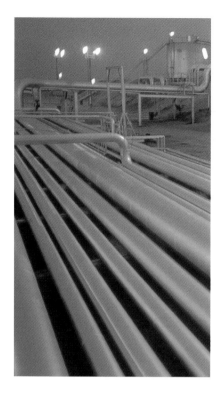

Bechtel was central not only to extinguishing the Kuwait fires but also to the full restoration of oil production and refining. Despite dire predictions, post-war output reached well over 200,000 barrels a day within a matter of months.

On August 2, 1990, Saddam Hussein's Iraqi troops invaded neighboring Kuwait, threatening to touch off a wider war in the Arabian Gulf. The conflict created uncertainty in international oil markets, setting off a worldwide economic slump. A host of new projects was put on hold.

Meanwhile, a drama bigger than any engineering project was unfolding in northern and central Iraq. Within days of the attack on Kuwait, Iraq took 109 Bechtel employees and their dependents hostage. In September, 10 Bechtel employees—three Americans and seven British nationals who had taken refuge in their respective embassies along with their colleagues still trapped in Iraq— were snared by Iraqi immigration authorities, who lured them out by insisting that they appear personally to secure exit visas for their wives. When they left the embassy compounds, these 10 were immediately swept up, taken to a facility on the outskirts of Baghdad, and held there. Meanwhile, their families and colleagues remained in their embassies, some camped outdoors, with winter approaching.

Riley Bechtel essentially camped out in his office for the duration. He had little choice but to keep a low profile and quietly organize activities to help the Bechtel employees held hostage. In the end, every Bechtel person was safely out of Iraq before the Allied Coalition launched its assault to liberate Kuwait. Bechtel could then savor the fact that a particularly determined (and fit) Bechtel manager had left the sanctuary of the British High Commission in early December, competed in a long-distance footrace through Baghdad, returned safely to the sanctuary, and was evacuated to the U.K. a week later.

THE AFTERMATH OF DESERT STORM

As the Desert Storm offensive took shape in nearby Saudi Arabia, Bechtel met quietly with Kuwaiti officials in London to lay plans for the restoration of their economic engine. Bechtel's three-person advance team landed in Kuwait early in March 1991, within days of the war's conclusion. The destruction that confronted them was beyond imagination. Before retreating, Iraqi troops had methodically devastated Kuwait's prized oil fields—750 wells were damaged, and

1991
_____

In the aftermath of the Gulf War, Bechtel leads the effort to put out 650 oil well fires and rebuild Kuwait's oil infrastructure, destroyed by Iraqi troops.

650 of those blazed ferociously. An estimated 70 million barrels of thick crude spewed onto the desert floor, forming lethal lakes.

Bechtel people had anticipated finding at least 200 sabotaged wells and severe damage to the refinery Bechtel had helped build 20 years earlier. Bechtel estimated it would take 4,500 workers to cope with reconstruction. The devastation was far worse than expected: no water, electricity, food, or usable facilities; nearly four times as many damaged wells as had been estimated; and deadly unexploded ordnance, some covered by the oil.

To borrow a phrase, Kuwait had become "hell with the lid off." Where fire didn't burst from the superheated ground with an ear-piercing whine, gushing oil did. Fire shot to the heavens in massive eddies of orange, red, and yellow flames. Black smoke blotted out the midday sun, filling the sky with a steady drizzle of soot mixed with tiny droplets of oil. The tacky coat of grime clung to vehicles, animals, and people.

Bechtel's assignment was to recruit, sustain, and manage the international force of 10,000 that was needed to extinguish the fires and restore petroleum production. Some experts estimated it would take as long as five years simply to control the fires. Under Bechtel's guidance, an army of workers did the job in eight months. Controlling the wells so quickly saved Kuwait Oil Company billions of dollars in irreplaceable reserves that otherwise would have gone up in smoke. More important, the people of Kuwait and neighboring countries were spared months, perhaps years, of additional severe air pollution, disruption, and deprivation.

Work on *al-awda* (the return) and *al-tameer* (the reconstruction) began in March 1991; the first postwar oil began to flow less than three months later. Kuwait came back on-line faster than anyone had thought possible, and oil production climbed steadily toward preinvasion levels. Over 1.5 million barrels of weathered crude were reclaimed from oil pits and lakes scattered across the desert. As work progressed ahead of schedule, it became clear that the initial estimates of $100 billion or more to rebuild war-ravaged Kuwait were, as the *Los Angeles Times* noted, "something of a Middle Eastern mirage." On January 12, 1992, just a year after the six-week war had begun, the *Times* ran a story headlined: "With Fires Out, Kuwaiti Business Bonanza Fizzles." The total cost of reconstruction was pegged at less than $5 billion, a figure that owed much to the skill, ingenuity, dedication, and courage of Bechtel's people.

1992

At Cape Canaveral, NASA's *Mars Observer* probe is launched from the Bechtel-built Launch Complex 40, the world's most sophisticated launch facility. The most complex structure on the site is a mobile service tower, designed to enclose the launch vehicle and then move away.

1992

A Bechtel joint venture is selected to provide design, procurement, and construction management services for a Motorola facility in Tianjin, China. The complex will produce a variety of electronic products and provide office space. Bechtel will work for Motorola in China for the rest of the decade.

One of the critical demands placed on managers in the 1990s was to identify what a company truly does best—its core competencies, in the jargon—and then focus its activities on them. There was never any doubt about Bechtel's core competencies or what the company's essential mission would be. "The engineering and construction business will always be our primary purpose," Riley told one interviewer. "Most conglomerates ultimately falter for lack of a central purpose or core business. . . . To me, designing and building projects is, and always will be, the spine of our business. We will never be a conglomerate. At least, not on my watch." Bechtel's essential mission was to be "the engineer-constructor of choice for customers, employees, and key suppliers in every industry market we serve."

In the early 1990s, Riley reemphasized the philosophy on which his father (in 1967) and his grandfather (in the 1950s) had formally anchored the company's success. This was not continuity for continuity's sake but a reaffirmation of what nearly a hundred years in business had proved time and again: customer focus, commitment, fair dealing, flexibility, private ownership, and a wide variety of finely honed skills distinguish Bechtel. These values and related concepts were articulated as:

OUR PURPOSE

Bechtel provides premier technical, management, and directly related services to develop, manage, engineer, build, and operate installations for our customers worldwide.

1993

Bechtel begins participation in Solar II, a prototype for a solar power plant efficient enough to compete with fossil fuels in the 21st century. Solar II, located in the Mojave Desert, is a follow-up to Solar I, which experimented with the storing of solar energy.

OUR MISSION

We will be the engineer-constructor of choice for customers, employees, and key suppliers in every industry market we serve by:
- delivering exceptional value to our customers;
- earning a fair return on our delivered value; and
- working closely with our customers, key suppliers, and communities to help improve the standard of living and quality of life.

OUR CORE PRINCIPLES

We bring to our work:

—a proud heritage of accomplishment, integrity, excellence, and commitment
to our customers' interests; and

—a willingness to appropriately adapt ourselves to change while maintaining
our fundamental values and constancy of purpose.

We will continue to:

—adhere to the highest standards of ethics and integrity;

—clearly and continuously understand, be totally responsive to, and strive to
exceed our customers' expectations; by our performance deliver exceptional
value to those customers, helping them to maximize their success;

—do our work safely and consistent with responsible environmental principles;

—make continuous improvement an integral part of the way we operate;

—attract, develop, motivate, and retain highly competent, committed, and
creative colleagues of diverse origins who are the best in their fields;

—promote from within as much and as early as possible;

—create a work environment, supported by leadership, that fosters openness,
trust, communication, teamwork, empowerment, innovation, and satisfaction;

—respond to our rapidly changing world with entrepreneurial approaches,
innovative solutions, advanced technology, and high-quality, timely decision making;

—remain privately owned, financially prudent, and global, with ownership
held by active senior management; and

—reward those contributing to our success.

1993

Bechtel is hired as part
of Navy CLEAN II
(Comprehensive Long-Term
Environmental Action
Navy), a project to clean up
the U.S. Navy's hazardous
waste sites.

To translate this mission into action, Bechtel's senior leadership formulated
seven umbrella strategies intended to spawn more specific initiatives within each
Bechtel organization. These guiding strategies were communicated to employees in
April 1992: perform; market and sell our special strengths; innovate to add value;
be the preferred E&C employer; be global; improve management and leadership;
understand and apply technology.

1993

The 800-mile Bechtel-
built PGT-PG&E natural
gas pipeline expansion is
completed.

The early 1990s saw an economy on the mend. By 1992, Bechtel revenues reached an eight-year high of $7.8 billion, the third straight year of steady growth. New work booked also enjoyed healthy gains, topping $9.4 billion in 1993. That was a 10-year high and nearly double the 1990 figure. Bechtel was at work on a host of high-profile projects in dozens of countries on six continents.

The Boston Central Artery/Tunnel project, the largest public works effort in the United States, included a freeway beneath the middle of Boston and a third vehicular tube beneath Boston Harbor to Logan Airport. When completed later in the decade, the Ted Williams Tunnel would earn the Outstanding Civil Engineering Achievement Award from the American Society of Civil Engineers.

In 1990, Bechtel Executive Vice President John Neerhout Jr. was named project chief executive of Eurotunnel, a consortium of five British and five French companies building the Channel Tunnel between England and France. Neerhout, with a very small team of Bechtel veterans, was charged with the task of turning around the enterprise's faltering engineering and construction effort. The $14.7 billion, 32-mile high-speed rail link was burdened with huge cost overruns and delays that were causing a breakdown in stakeholder confidence. It was an enormously intricate project involving two languages, two governments, three national railways, numerous subcontractors, and a syndicate of 220 banks. Bechtel helped manage the project to its successful completion in 1994, providing management, technical, and construction expertise, which helped restore the trust of investors and financial institutions.

Also in 1990, Bechtel's skills were called on to help manage Hong Kong's US$20 billion program to build a new international airport and related infrastructure. The project was the largest civil infrastructure program in the world. It comprised not only the airport, built on a largely man-made, 3,100-acre island, but rail and road systems linking the island to the mainland and a planned community for 20,000 residents. The Hong Kong Airport Core Programme also included the world's longest suspension bridge to carry both road and rail traffic, as well as a 1.2-mile tunnel under Victoria Harbour.

On the other side of Asia, Bechtel and Turkish partner Enka designed and built a 140-mile segment of Turkey's transnational highway system. Bechtel

Some 2,000 movable mirrors reflected sunlight up to a receiving tower at the Solar II project. This solar energy was converted into enough electricity for thousands of California homes.

Enterprises arranged several rounds of financing for the undertaking, part of the $1.4 billion Trans-Turkish Motorway linking Europe and Asia.

Bechtel also was busy on seven new airports besides Hong Kong, from Dubai to Dallas. At Daya Bay, Bechtel helped build the first commercial nuclear power plant in the People's Republic of China. In fact, Bechtel would become the first U.S. company ever granted a construction license in China (as well as in Japan and South Korea during the same period). In Venezuela, the company won an engineering, procurement, and construction contract for the process units at Maraven's Cardon refinery. In Kazakhstan, Bechtel and Enka won a contract to help develop the Tengiz and Korolev oil fields. At scores of sites in more than a dozen states, Bechtel managed major environmental cleanups for the U.S. government. In Papua New Guinea, Bechtel construction management helped Chevron extract oil in the Southern Highlands. Fifty miles off the coast of Qatar, Bechtel and Technip completed their managing contractor role helping Qatar General Petroleum Co. tap one of the world's largest remaining reserves of natural gas. Bechtel finished up program and construction management help to The London Docklands Development Corp. in transforming a decrepit waterfront area on the Thames into a third city center. And there were new awards for gas fields in Abu Dhabi, pipelines

As part of the Cooperative Threat Reduction program, Bechtel helps Russia develop a comprehensive chemical weapons destruction plan.

1995

Bechtel introduces PowerLine® to offer potential owners already-engineered, standard power plants with world-class operating results at low cost on aggressive schedules. The first PowerLine project completed is the Hermiston generating plant in northeast Oregon.

OPPOSITE: *To nearly double the amount of natural gas delivered from Alberta, Canada, to central California, Bechtel expanded the PGT-PG&E natural gas pipeline using some 400,000 tons of pipe. Bechtel built the original line in 1961.*

in Algeria and Thailand, and a Motorola factory in China that would produce pagers, semiconductors, and cellular telephones.

But for all the company's prominence, Bechtel's management team recognized the need for profound change. Its members were far from convinced that the company was functioning as effectively and efficiently as it could or would need to be.

PROCESS IMPROVEMENT AND REENGINEERING

For decades, the company's focus on improving performance centered on what Riley Bechtel summarized as "dedication to excellent results, management of quality, and quality of management." It was time for a more rigorous, formal approach.

In mid-1992, Bechtel leadership supplemented its continuous improvement model with a corporate business model comprising seven critical processes: satisfying customer needs; finding and defining opportunity; developing opportunity and customers; performing work; developing and applying technology; planning and allocating material resources; and building teams. Led by "home-grown" continuous improvement coaches, employees formed teams across the company to formalize and clarify what had come naturally to some but was still unclear to many.

The essence of continuous improvement is dissecting a process, improving it, measuring progress, and so on—with no end. But the nature of continuous improvement is that it yields only incremental gains. And, as Riley would note, this inherently inch-by-inch approach would in itself not be enough. In some places, a sea change was needed. "Reengineering is more than continuous improvement. It isn't about fixing something, it's about leaping forward," observed Riley. Bechtel, he said, had to fundamentally rethink and redesign its work processes without destroying the underlying principles and values that had made the company great. Intensive education about eliminating corporate waste helped crystallize the notion that breakthrough improvements were not only attainable but essential. Bechtel people enhanced their ability to cut costs, accelerate schedules, and maximize customer satisfaction.

Bechtel management set an ambitious goal: reduce the total installed

Bechtel and PG&E expand their successful relationship by forming International Generating Co. (InterGen) to develop, own, and operate electric power facilities outside North America.

Under the U.S. Department of Defense Cooperative Threat Reduction Program, Bechtel has helped nations of the former Soviet Union dismantle their nuclear and chemical weapons. In Ukraine, the company has coordinated the destruction of 160 nuclear missile silos.

capital cost of a typical Bechtel-designed and -built facility by 30 percent, reduce schedule requirements by 30 percent, and create a 30 percent increase in value delivered. And do all of this in three years. This became an initiative known as Engineering, Procurement, and Construction Process Innovation.

Leadership was tasked and resources allocated. Bechtel focused on the power and petroleum-and-chemical industries, sectors best suited to the kind of replication helpful to meeting these aggressive goals. The process began with the conception and layout of a plant and extended, in some cases, all the way through to maintenance and operation. Bechtel's work processes were enhanced and automated methods were introduced. By 1998, Engineering, Procurement, and Construction Process Innovation had become embedded in Bechtel's project methodologies.

IMPROVED MARKETING, BUSINESS DEVELOPMENT, AND STRATEGIC SKILLS

By the mid-1990s, it was increasingly clear that Bechtel's continuous improvement and reengineering initiatives should be complemented by better marketing and strategic skills.

A special task force of marketing leaders from throughout the company convened in 1995. They combined their understanding of Bechtel's business with their continuous improvement and reengineering knowledge to produce in two months a blueprint for sharpening the company's marketing and business development efforts.

Bechtel had shown over the years, perhaps never as dramatically as in Kuwait from 1991 to 1993, that it could respond to customers' needs faster and better than any other engineer-constructor. The company now needed to improve its ability to decide much earlier how, when, and where to place its bets—in terms of human and financial resources.

Efforts included conducting extensive strategic assessments of local markets thought to be attractive; making planning and budgeting processes more dynamic, substantive, and disciplined; and promoting longer-term relationships with customers, prospects, and partners.

CORPORATE CHANGE

It became increasingly clear that the perception of Bechtel in many parts of the world was changing—and not always for the better. Bechtel, some thought, was a fair-weather player, maintaining a market presence only when opportunity abounded and pulling out during lulls. Bechtel didn't sufficiently understand many local markets and important relationships. Bechtel wasn't maximizing the use of local resources to the benefit of its customers and itself. And some found the company arrogant.

Bechtel had entered the 1990s with an organization centered on industry-oriented business lines. Since the 1960s, the company had been structured in matrix fashion with an industry axis and, secondarily, a geographic axis. The business lines made all the decisions on where and how to allocate company

Every year, some 100,000 tons of recycled pulp are turned out at the recycling facilities Bechtel built for FSC Paper Co. in Illinois. Environmental awareness is a keynote of the 1990s.

resources. They bore responsibility for profit and loss.

Riley wanted to create a flexible organization that could respond more quickly and effectively to fast-changing market forces. By mid-1994, the senior executive team was convinced that a more balanced matrix—built on a decentralized management structure—would make Bechtel more competitive and more responsive to customers' needs. The global business lines and regional offices would now share decision-making responsibility and authority for project management and performance. Project managers reported into both sides of the matrix. Further, either side could serve as a customer's "window into Bechtel," depending on that customer's own preference.

By the end of 1995, after extensive debate and careful analysis, senior management decided to refine the matrix further and make some attendant changes.

"Regions," modeled on the Europe, Africa, Middle East & Southwest Asia operation formed in 1991 and led from London by executive committee member Don Gunther, were given responsibility for profit and loss from winning work through performing it. Project managers reported squarely to the regional leadership. The new regions included the Americas, which would later be split into North America and Latin America, and Asia Pacific. Some regions would create subregions.

"Global industry units" were chartered to create value for delivery by a region. These GIUs (including Power, Petroleum & Chemical, Mining & Metals, Civil, Industrial, and Telecommunications) were charged with maintaining industry specialist capabilities (many of which are dynamically deployed to a region for winning and executing work), leading global industry relationships, and developing world-class conceptual and basic designs.

Bechtel Systems and Infrastructure, a combined region and GIU, was to serve the U.S. government, and regional, state, and local governments where substantial federal funding was involved.

The center of Bechtel's decision-making was now physically closer to the company's customers. The new organization would also give Bechtel much more local knowledge and presence. And accountability for integrated decisions (addressing benefits and costs, short- and long-term considerations, etc.) would be more clearly defined and promote better decisions.

This "region-tilted" approach was announced and implemented in early

Cows graze placidly against a background of one of two petrochemical plants built in the mid-1980s by Bechtel for Tennessee Eastman several thousand miles apart—one in England and one in South Carolina—using a single design that is as versatile as the plastic resin the plants produce. From that resin come products as varied as soda bottles and yarn.

OPPOSITE: *Icebreakers plowed through frozen waters to reach northern Quebec just below the Arctic Circle—one of the most unforgiving places on Earth. Bechtel provided engineering, procurement, and construction management services for the Raglan nickel and copper mine project there.*

James Balog, Tony Stone Images

Despite difficult conditions, Bechtel finishes Algeria's portion of the 335-mile Maghreb-Europe natural gas pipeline four months ahead of schedule. The pipeline runs from Algeria's main gas field in the Sahara across northern Morocco and under the Strait of Gibraltar to Córdoba, Spain, where it ties into the Spanish gas transmission network.

Bechtel is selected as the prime contractor for Atlantic LNG, the Western Hemisphere's first liquefied natural gas plant in 25 years. The plant, in Trinidad, becomes the largest single investment ever in the Caribbean.

Bechtel Water Technology creates the world's largest microfiltration water treatment facility in the United Kingdom. This project for North West Water is built under a tight deadline to help the drought-stricken residents of Merseyside.

1996. Riley's announcement also included, as he would later put it, "retirement with full military honors" of Bechtel's Executive Committee concept. The functions of this decision-making body, typically composed of line executives reporting to the CEO, were supplanted mainly by improved delineation of roles among senior executives.

Riley would chair an informal group called the Chairman's Leadership Council, which would consult periodically on key decisions. Serving on the council with Riley would be Adrian Zaccaria, president and chief operating officer; Vice Chairman Fred Gluck (who retired from line duties in mid-1997); Don Gunther (who became vice chairman in July 1997 and retired from that position in July 1998); and Executive Vice President John Carter. Chairman Emeritus Steve Bechtel Jr. would be a permanently invited guest.

Though successful at this writing, this structure wasn't cast in concrete. "When circumstances have changed—especially market needs—we have never hesitated to adjust our organization," noted Riley.

BECHTEL LEADERSHIP, COVENANTS

Of all the important management initiatives of the turbulent 1990s, Riley saw none more important than leadership. "More than anything else, our interrelated initiatives, from reengineering to restructuring the company, all depend on improved leadership for success."

Quality leadership had always been the subject of management attention over the decades at Bechtel, but Riley expected the complexity and competitiveness of the global economy to place new demands on leaders in the coming millennium.

In late 1992, Bechtel established a special continuous improvement team, with leaders drawn from across Bechtel management lines under the chairmanship of Executive Vice President Don Gunther, to examine the challenges of building teams with particular emphasis on developing a formal model of the ideal Bechtel leader.

The resulting detailed model, designated "2001 Leadership" and later referred to simply as "Bechtel Leadership," would promote the development of leaders who focus on their customers, build teams, empower their colleagues, and

QA Photos Ltd

practice continuous improvement. Bechtel Leadership, as written, practiced, and continually evaluated, is a system of individual attributes and judgment qualities (what Bechtel leaders try to *be*—technically competent, resilient, persistent, etc.) combined with skills and competencies (what Bechtel leaders try to *know*—coaching, problem-solving, organizing, etc.), and leadership work processes (what Bechtel leaders try to *do*—understand and define the purpose and mission; clarify Bechtel's values and core principles; create a shared vision; develop goals and objectives; develop plans to attain goals and objectives; and implement, monitor, and improve the process).

Laser-guided tunnel boring machines were used to excavate the Channel Tunnel. The mammoth machines chewed their way through the soft chalk deep beneath the English Channel.

The Rocksavage plant will be the most efficient power plant in England, if not the world. InterGen owns and will operate Rocksavage; Bechtel is building the Merseyside plant and providing technical support.

1996

Bechtel begins work on an $800 million, 440-megawatt power plant in Quezon Province in the Philippines. Bechtel companies provide project development and finance, project management, engineering, procurement, and construction services, while InterGen serves as the project's managing partner.

To promote the practice of Bechtel Leadership throughout the organization, the first Bechtel Conference of some 300 leaders from across the company in June 1997 adopted the following "Leadership Covenants" to guide everyone at Bechtel:

1. Treat Bechtel colleagues with mutual respect, trust, and dignity and believe they are acting in the best interest of the company.

2. Help each other; ask for and give help and welcome it freely (it is not a sign of weakness). Go out of the way to provide extra support to fellow employees. Share experiences and lessons learned, both successes and failures.

3. Communicate early, honestly, and completely with all who have a direct interest in the subject. Listen to others' points of view.

4. Earn trust by accepting and honoring agreements, keeping promises, and discussing needed changes before acting.

5. Work to understand Bechtel Group, Inc. goals and strategies and proactively support them through discussions, communications, and actions (for example, sharing resources).

6. Never undermine colleagues directly or indirectly.

7. Work jointly to resolve disagreements in good faith. If necessary, go to a higher authority together, then accept and support the solution.

8. Contribute constructively by exercising the highest level of professional and ethical behavior.

9. Promote continuous use of the covenants.

PRIVATIZATION AND PARTNERSHIPS

In a world increasingly long on infrastructure needs and short on private capital and government funding, Bechtel's financial leverage and entrepreneurial touch have proved to be powerful competitive tools.

Bechtel Enterprises grew out of Bechtel Financing Services, Inc., which had been formed in 1969 to help customers with project financing by assembling international export credits and commercial bank loans. Bechtel Enterprises would do all this as well as develop projects and take equity positions, creating a number of joint-venture companies, starting in the power sector. These capabilities help differentiate Bechtel from its competition, provide innovative world-class ownership teams, and create attractive investment opportunities for Bechtel and its partners.

Bechtel had been involved in independent power generation for some time when in 1989 Bechtel Enterprises formed a strategic partnership with PG&E Enterprises called U.S. Generating Company. When the U.S. Energy Policy Act of 1992 made it easier to build and finance independent power plants, USGen took off, successfully developing more privatized power plant capacity than any other independent power company in the country. InterGen, another joint venture with PG&E, developed, owned, and operated electrical generating facilities outside North America, starting in 1995. As power privatization expanded globally, work got under way in the United Kingdom, the Philippines, Mexico, and Colombia.

In 1997, the two companies agreed to specialize and separate their progeny, with Bechtel buying PG&E's share of InterGen and PG&E buying Bechtel's share of USGen. Bechtel later sold 50 percent of InterGen to Shell. Among InterGen's projects is Quezon Power, the first large-scale independent power project in the Philippines to be financed, built, owned, and operated by a private entity without requiring the sovereign backing of the national government.

Meanwhile, Bechtel Enterprises and its partners branched out into privatized water, transportation systems, and other infrastructure ventures. A few examples illustrate the global reach and local touch of Bechtel Enterprises. Joint ventures with local, regional, and international partners are working to:

– finance and build in China a six-lane road linking industrial areas on the Taiwan Strait and the Pearl River Delta

– expand and operate water and wastewater services for Manila

– lease, upgrade, and operate Perth International Airport

– help finance, build, and operate Mexico's first privately owned natural gas pipeline, supplying fuel to the Yucatán Peninsula

– build, own, and operate two wastewater treatment plants in the Scottish Highlands

– help finance and build the Dabhol power plant near Mumbai (Bombay).

Bechtel continues its practice of partnering locally. For example, in

1996

Bechtel begins construction on the 700-megawatt Samalayuca II project. The $647 million plant is the first privately funded power project in Mexico and is owned by an international consortium that includes a Bechtel Enterprises affiliate.

1985 Bechtel and China International Trust and Investment Co. (CITIC), the state-owned entity that is China's largest investment company, established China American International Engineering, Inc. (CAIEI), a joint venture for appropriate engineering and construction in China. During the 1990s, CAIEI built Motorola's world-class electronics complex in Tianjin, PPG's float glass plant in Dalian, and other projects. Bechtel and CITIC also formed a joint venture to develop a new superport and associated infrastructure south of Shanghai on Daxie Island.

BEYOND 1998

"What may be unique about Bechtel at 100," said Riley, "is the extraordinary bond of commitment, competence, and integrity that we've been able to forge—with customers, with key partners and suppliers, and with each other as colleagues in pursuit of great accomplishments." That bond has helped Bechtel thrive in a punishing industry, in which few companies survive more than half a century.

"We know that continuing to lead this industry for another century requires the will and the skill to be the best of the best," Riley said. "That's the only way to succeed on our scale and at our level in a world of accelerating change and rising expectations. To lead its industry for another hundred years, any company operating today will have to be exponentially better than it already is."

Bechtel plans to be one of them.

Fiber-optic cable and digital signals are shaping tomorrow's telecommunications networks, but callers from Cleveland to Medina want it all yesterday. Bechtel is on the line, speeding up installation and controlling costs. By 2001, Saudi Arabia will have 2 million new lines, 500,000 of them wireless.

1998

Hong Kong International Airport is dedicated.

OPPOSITE: *The Hong Kong Airport Core Programme is one of the world's largest infrastructure projects.*

PART II:
PROJECTS

HOOVER DAM
Nevada & Arizona, 1931–36

When the U.S. Bureau of Reclamation's plan to construct Hoover Dam was first announced, it was not without its detractors: Too much concrete, some said; the heat generated by the setting concrete would crack the dam, said others; some predicted the reservoir would silt up rapidly and the dam would become useless.

But for the farmers of California's Imperial Valley, Hoover Dam was good news. It would mean an end to the unpredictable floods rising from the Colorado River; its harnessed waters could be used to irrigate one of the potentially richest agricultural regions in the world; and the output from Hoover's massive 1,350-megawatt power station would be divided among Nevada, Arizona, and Southern California municipalities.

For members of Six Companies, including Bechtel, Hoover represented the challenge of a lifetime. They put up a $5 million bond as security, committing them to fulfill the $48.9 million contract. It was the largest civil engineering project in the country's history and a dam that would control one of the most powerful rivers in the world.

The Hoover project placed huge logistical demands on Six Companies: 3.7 million cubic yards of rock to excavate, 45 million pounds of pipe and steel to erect, 4.4 million cubic yards of concrete to pour. The dam is a building-block set of interlocking concrete modules that vary in size from about 60 feet square at the upstream face of the dam to about 25 feet square at the down-

Water from Lake Mead, which was created by the dam, is fed to Hoover's turbines through four 395-foot-high intake towers, two on each side of the canyon. Here, before the lake was at capacity, the towers are visible.

Hoover Dam stands 726 feet high, about 1,200 feet across the crest, and 660 feet thick at the base. At its peak in June 1934, the project employed a workforce of up to 5,000 at a time, which worked a three-shift day, around the clock, seven days a week, to keep to the schedule. The first task was to divert the Colorado River's unpredictable flow, which can vary from 3,000 to 300,000 cubic feet per second. A special jumbo rig that could move two dozen drillers at a time (below) was used to dig four cavernous diversion tunnels, each 56 feet in diameter. The tunnels were roughed into the canyon rock to reroute water away from the site (right).

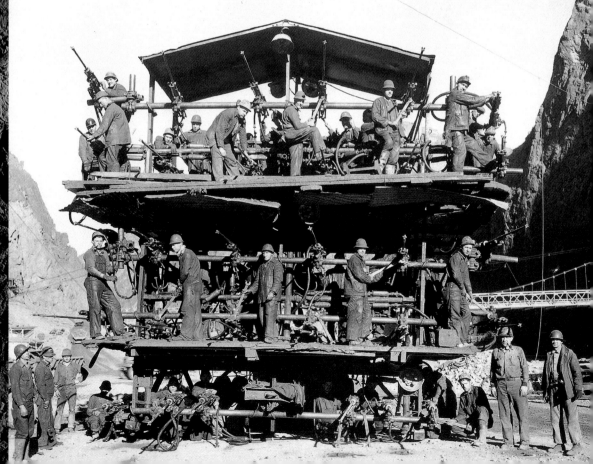

Bureau of Reclamation

stream face. To dissipate the tremendous heat being released by so much setting concrete and to speed the process, more than 582 miles of one-inch steel pipe was embedded into the blocks, and cold water from a refrigeration plant circulated through them to accelerate curing.

Hoover Dam not only held back the Colorado River, it also held back the Depression for the army of men who labored there. Thousands flocked to the construction site looking for work. They found jobs, and an extremely challenging environment. Temperatures in the summer months could reach 120 degrees Fahrenheit, and flooding was a constant threat. But somehow the challenge of building something so vast, so important, and so elemental made it bearable; these men knew they were part of history.

Remarkably, the entire project was completed under budget and two years ahead of schedule. Although Bechtel has since built bigger dams, none will ever be quite as important, to either the company or the country. It was Bechtel's first mega-project and it shaped the company's ambitions forever. After Hoover, Bechtel would always be able to think big.

An overhead cableway was installed to transport men and materials across the chasm (opposite). Before construction crews arrived at the Black Canyon site in Nevada, there was little but high desert and towering rock. Workers built an entire town, including stores, schools, and even a rough golf course, football field, and Hoover-scale pool parlor (below) for the thousands of people employed to build the dam. Within months of the start of construction, Boulder City became the second-largest town in Nevada.

Bureau of Reclamation

SAN FRANCISCO–OAKLAND BAY BRIDGE

California, 1933–36

The trickiest part of the bridge was the piers. Contending with strong tides and winds, steel caissons were sunk with great precision to allow people to work deep beneath the bay (below). To complete the job, Bridge Builders painted 70,815 miles of cable (opposite).

For decades, Bay Area residents had dreamed of a bridge linking Oakland with San Francisco. On November 12, 1936, their wish came true. The San Francisco–Oakland Bay Bridge, at 8.2 miles, remains one of the longest in the world and lays claim to being the world's busiest toll bridge, carrying more than 270,000 vehicles each day.

The bridge was another of the great public works of the Depression, costing some $78 million. San Franciscans marveled as it went up in two pieces, linking at Yerba Buena Island. Yet some of Bechtel's best work isn't visible. As part of Bridge Builders, Inc., Bechtel was responsible for the pier foundations of the east bay crossing's cantilever span and faced the challenge of the bay's powerful currents. The piers had to be set deeper than any previously attempted, setting a world record at minus 247 feet.

The trickiest part of the bridge was the piers. Contending with strong tides and winds, steel caissons were sunk with great precision to allow people to work deep beneath the bay (below). To complete the job, Bridge Builders painted 70,815 miles of cable (opposite).

CALSHIP AND MARINSHIP
California, 1941–46

Marinship's and Calship's achievements are all the more impressive given the dearth of skilled labor available. Less than 10 percent of the workforce had ever been employed in a shipyard. With experienced shipwrights scarce, managers lowered the skill requirements by breaking the work into subassemblies (opposite).

Shipbuilding is supposed to be an art form, requiring the different skills of scores of workers and plenty of time. But at Calship and Marinship, hulking ships were welded together in sequences like giant three-dimensional steel puzzles. They had to be. There was a war on, there was neither time nor skilled labor, and the Allies needed ships in vast quantities. Bechtel-McCone–sponsored war production companies, Calship in Los Angeles and Marinship north of San Francisco, pulled off a manufacturing miracle during World War II. Operating nonstop throughout the war, the two yards broke records for shipbuilding speed, constructing 560 ships in the time normally required to build one-fifth as many. Liberty and Victory cargo ships, T-2 tankers, and oilers came rolling out of the yards. On average, Calship was delivering a new vessel every 65 hours. Yard construction at Marinship in Sausalito began on March 28, 1942, and its first Liberty ship was launched on September 26 of that year.

The war forced managers to take an entirely different approach to manufacturing. Of the more than 60,000 workers who toiled at the two yards during the peak of activity, the vast majority had no experience as shipwrights. Creating a Liberty ship meant bringing together a quarter-million separate items in a series of carefully orchestrated steps. Instead of constructing ships from the keel up, builders broke down the task into a series of subassemblies. Workers could then be taught the required skills needed to complete one section of a ship. At Marinship, final assembly brought together 108 units—keel sections, deck sections, deck houses—into a massive tanker or oiler. The

relentless pressure on production cut by nearly 80 percent the time it took to produce such a ship.

While functioning as round-the-clock, full-bore wartime factories, the shipyards became a crucible for cultural change. Marinship could boast that it had the highest percentage of female workers of any yard in the nation. The breakthrough for women craftworkers came in the summer of 1942 when Marinship hired Dorothy Gimblett, a mother of three, as a welder. Her appearance in heavy welder's leathers was a major event, but women quickly proved themselves to be the equal of men in welding and many other jobs. Minority workers also gained access to skilled jobs in unprecedented numbers, working alongside whites and earning exactly the same pay and benefits.

"Man" power was so short that women stepped into even the grittier jobs, such as welding (above). Calship's facility (opposite) included 14 shipbuilding ways and 10 outfitting berths.

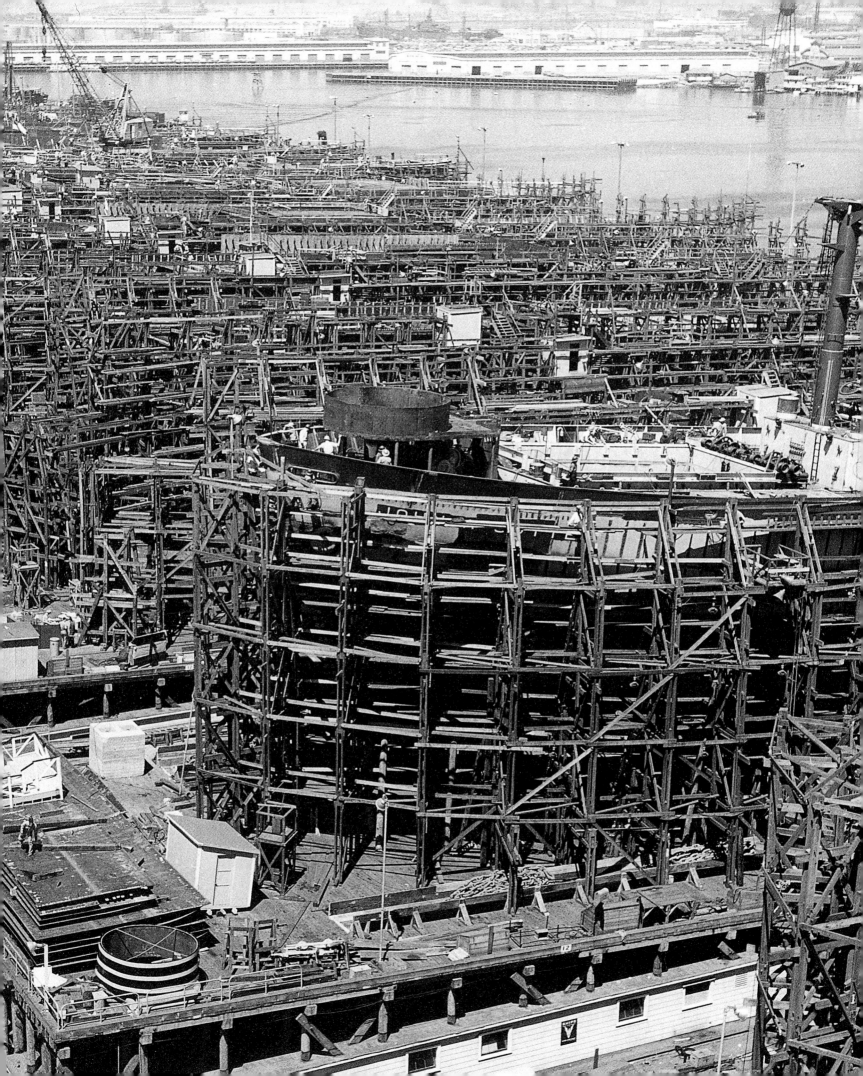

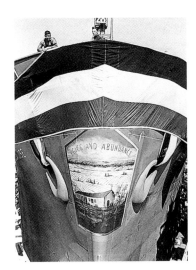

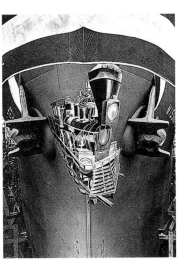

*Those responsible for Calship are proud
of its record. Year after year, operating
three shifts a day, seven days a week, we
produced ships in unprecedented numbers
and at low cost.*

JOHN McCONE

PRESIDENT, CALIFORNIA SHIPBUILDING CORPORATION

TRANS-ARABIAN PIPELINE
Saudi Arabia, 1947

Stretching like an enormous straw from the Abqaiq oil field to the Mediterranean port of Sidon, the 1,068-mile Trans-Arabian pipeline allowed Europe to quench its growing thirst for oil by drawing on the deep pools of the Middle East. Between those two terminals was an expanse of desert that some American oil drillers called *wajid niafi*, Arabic for "plenty of nothing." That's not quite true. There were plenty of things in the Arabian desert: scorpions, flies, and raging sandstorms called *shemals* that could last for days. Not a single permanent structure existed along the Tapline route: no road, no building, no well.

Although the desert presented problems, its openness allowed engineers to attack the mission on a large scale. Tapline was the mightiest pipeline ever constructed, a "big-inch" conduit, 30 inches in diameter, which initially carried some 300,000 barrels of oil a day, eliminating the need to ship crude oil by tanker. At the eastern construction camp of Ras al Misha'ab, a three-mile overhead cableway transported loads of 31-foot pipe sections from freighters docked in the Arabian Gulf. The pipes were welded into 93-foot lengths. Caravans of trucks that had been modified for desert work hauled 40-ton pipe loads to the line. There, side-boom tractors—equipment that had been developed by Bechtel—put the pipe in place, and the welders went to work.

The logistics were impressive. The work crew, 14,500 Arabs and 1,500 expatriates at its peak, moved 2.5 million cubic yards of sand and rock, built 1,200 miles of access roads, and handled 265,000 tons of steel pipe. About half the line was built above-ground, resting on cement pads or steel bents. For subterranean sections, a giant 20-ton ripper with a six-foot "tooth" gouged out trenches in the limestone that underlay the sand in some parts.

The construction crew members who came to work in the desert referred to themselves as "Camel Legionnaires." They were tough, hardworking, and proud. Carefully screened and well paid, they were the best craftsmen in the business.

Tapline was very important for Bechtel, which built 80 percent of the line with joint-venture partners. The pipeline represented a step forward in King Abdul Aziz Bin Saud's ambitious plans for his country, and cemented a relationship that would lead to huge projects such as Jubail, an entire city built in the desert.

TRANS MOUNTAIN PIPELINE
Canada, 1952–53

Bechtel pipeliners drove the sweeping 700-mile, 24-inch Trans Mountain pipeline through the Canadian Rockies to bring Alberta's much-needed crude oil to the West Coast. There, it would be processed by refineries, freeing the Pacific Northwest of dependence on oil tankers.

For Bechtel as well, the pipeline would provide new independence, as it was the first major job in which the company served as engineer, project manager, and part owner, setting a critical precedent for the future.

The line ran through forests and high deserts, rose again to meet the Coast Range, then plunged 3,600 feet through canyons, before following the Fraser River to Vancouver.

Mindful of the land's ecology, a pleased Steve Bechtel Jr. recalled, "When we were all finished and cleaned up, you could barely see where the pipeline went through."

As many as six separate crews laid an average of three miles of pipe a day during two frantic summers of building the Trans Mountain pipeline from Edmonton, Alberta, to Vancouver, British Columbia. Trans Mountain was the first pipeline to cross the Rockies, and crews had to deal with topography that varied from steep canyon to boggy muskeg. Among the obstacles were 56 highway crossings and 72 rivers and streams. Traversing the Coquihalla Canyon (opposite) was one of the biggest challenges.

BAY AREA RAPID TRANSIT (BART) SYSTEM
California, 1959–76

The San Francisco Bay Area Rapid Transit (BART) system took the 19th-century concept of mass transportation and vaulted it into the 20th century. BART was among the first systems designed to be fully automated and computer-controlled, right down to fare collection. Planning for its aboveground portions included landscape architecture and right-of-way beautification to ensure that the railroad would be aesthetically pleasing.

BART was designed and built by a joint venture of Parsons Brinckerhoff, Tudor Engineering, and Bechtel. Bechtel's work, which received a dozen major engineering awards, included the transbay tube, built in 330-foot-long sections that would rest in a trench at the bottom of the bay. The 3.6-mile tube required exacting design and seismic studies before construction could begin. Before the rails went down, walking or biking the finished tube was a popular urban adventure for Bay Area residents. Bechtel's excavations for downtown subway stations were an adventure of another sort. Working 80 to 100 feet below ground, sandhogs operated under compressed-air conditions to hold back the groundwater. They had to work around a hundred years' worth of underground infrastructure. The dig in lower Market Street, an area that had once been part of the harbor, was rich with buried ships and other artifacts, providing a glimpse of an earlier San Francisco.

BART helped ease congestion on the Bay Area's notoriously busy highways. Dramatizing the impact of high-quality infrastructure, the extensive BART system has transformed the daily lives of Bay Area residents. Dozens of new communities have sprung up in the counties served by its train lines.

Every weekday, the BART
system carries about 270,000
passengers, and it has one of
the best safety records in the
world. On a passenger-mile
basis, BART's use of energy
during rush hours in the
commute direction is about
10 times more efficient than
that of an automobile.

SUNCOR ATHABASCA TAR SANDS PROJECT
Canada, 1962–67

The prehistoric muck known as the Athabasca tar sands contained a treasure that Canada had longed to tap. The sparsely settled area of northern Alberta was 12,000 square miles of oil-filled sand called siliceous marlstone, up to 300 feet deep, which held some 250 million barrels of recoverable oil. The problem was that no one had figured out an economical way to separate the oil from the sand.

Canadian Bechtel Ltd. cracked the code for Great Canadian Oil Sands Ltd. (later called Suncor), excavating a gargantuan open-pit mine capable of processing the sand in enormous volumes. It works like this: Two huge bucket-wheel excavators gobble more than 100,000 tons per day of oil-soaked sand. At the plant, hot water and steam are used to separate the sticky oil, called bitumen, from the sand and other solids. The bitumen is upgraded by a fluidized-bed coking process, then treated with hydrogen to remove impurities, producing high-grade synthetic crude oil. The crude is pumped through a Bechtel-built pipeline 250 miles south to Edmonton. To close the loop, the coke produced from the bitumen is used to fuel the generators that keep the whole works going.

CHURCHILL FALLS HYDROELECTRIC PROJECT
Canada, 1966–74

Nature saved some of its best work for Canada, a country of incredible beauty and enormous scale. It has water flowing in untold quantities from thousands of lakes and rivers carved out by glaciers, many of them far to the north of population centers. That water represents raw hydroelectric power, and Bechtel helped harness this power. At Churchill Falls, on the glacial plateaus of Labrador (700 miles northeast of Montreal), Bechtel managed what was then the largest single-site hydroelectric project in the world.

Churchill Falls presented the kinds of obstacles that Bechtel Canada engineers were used to and was located in a remote area with no roads or infrastructure. The joint venture Acres Canadian Bechtel (ACB) had to build structures to impound a watershed of 26,000 square miles and channel trillions of gallons of water through an underground powerhouse that had been hewn from solid rock.

Six huge concrete spillways and control structures and 40 miles of dikes and dams control the water flow through the Churchill Falls powerhouse.

The joint-venture team employed 6,200 men to excavate and move massive amounts of earth and rock, about 2.3 million yards for the powerhouse alone. Eleven generating units each produce 475,000 kilowatts of power. Churchill Falls was one of the most impressive technological achievements of its time.

Ted Levin

JAMES BAY HYDROELECTRIC PROJECT
Canada, 1972–85

Bechtel once again found itself working on a magnificent scale in the Canadian wilderness, helping manage James Bay hydroelectric project for Hydro-Quebec. As in the earlier but smaller Churchill Falls project, James Bay meant harnessing the power of a number of waterways, in this case a 68,000-square-mile hydrographic basin (bigger than the state of New York), known appropriately as La Grande Complex.

The "grand" in that name is the La Grande River, which flows 535 miles west into James Bay, after plunging 1,245 feet. However, by diverting neighboring Caniapiscau, Laforge, Eastmain, and Opinaca rivers, Bechtel engineers were able to nearly double the watershed's power potential. The project required 203 million cubic yards of fill, 138,000 tons of steel, 550,000 tons of cement, nearly 70,000 tons of explosives, and an enormous amount of determination. As many as 12,000 workers, many laboring 60-hour weeks, built 215 dikes and dams and constructed 4 power-houses capable of generating 10,300 megawatts. Getting that power south to Quebec and Montreal required building a 3,000-mile network of 735-kilovolt transmission lines.

The sprawling James Bay complex (below) harnesses the energy of four rivers and involved the construction of four powerhouses and 215 dikes and dams. The largest of four power stations, La Grande 2, has a capacity of more than 5,000 megawatts.

NORTH SEA OIL
Scotland & Norway, 1972–Present

There are billions of barrels of oil sitting below the North Sea, but this tempestuous body of water doesn't give up its bounty without a fight. It's a mariners' graveyard, famous for mean weather and 50-foot swells. For oil drillers, there's an added complication—the North Sea is deep, 500 feet in many locations. Bechtel helped pioneer offshore oil development in the North Sea beginning in 1972, converting a semisubmersible production platform for use in the Argyll field for Hamilton Brothers, which produced the first commercial quantities of North Sea crude. Since that time, Bechtel has participated in a number of other firsts and provided a variety of technology and management services to major North Sea oil companies, including Conoco, Gulf Oil, Mobil, Occidental, Sun Oil, and Statoil, Norway's government-owned oil producer.

For Conoco and others, Bechtel was project service contractor for the construction of a jacket for its Murchison platform. The Murchison jacket is a 26,000-ton steel frame that is 866 feet tall, slightly taller than New York's 70-story GE Building in Rockefeller Center. The jacket was built on its side and then towed to the location and sunk in place in water 515 feet deep.

Conoco's floating tension-leg platform was assembled in sheltered waters (above left), then towed to its site and tethered at each corner to templates (below left) fixed to the sea floor. The Piper platform (opposite) was launched from a barge.

For Statoil, Bechtel made use of a Norwegian-designed gravity-base platform. As the name implies, gravity-base structures are not anchored to the bottom but use their considerable mass to stay in one place. Bechtel managed construction of the Gullfaks "A" platform, the company's first major platform in the Norwegian sector. The monster's total weight is about 700,000

tons; it rises 709 feet from the seafloor. The base is formed by a honeycomb of 24 concrete cells, each measuring nearly 76 feet in diameter and 231 feet in height. Four of the cells are extended upward 300 feet, to act as supports for the steel deck frame that holds the drilling and production facilities, as well as living quarters. Most of the cells have a function other than weight: They store crude oil and diesel fuel, which also help ballast the platform. One cell is used to pump gas, oil, and water to the surface, where processing takes place.

The massive 16,000-ton jacket (right) for the Beryl "B" platform was towed to its site and sunk in 400 feet of water, with a tolerance of just 3.5 feet of the targeted position (above).

For Conoco, operator of the Hutton field, Bechtel provided engineering support for the first tension-leg platform. Unlike conventional platforms, which are anchored to the sea bottom and immovable, a tension-leg platform floats on the surface. It is tethered to foundation templates on the seafloor by tubular steel mooring lines, or legs. The buoyancy of the platforms creates an upward force, keeping the legs under constant tension. And given the huge cost of construction, tension-leg platforms have the added attraction of being reusable—the legs can be disconnected and the platform towed to the next location.

SYNCRUDE ATHABASCA TAR SANDS PROJECT

Canada, 1973–78

Syncrude's production process is similar to that used in the Suncor Athabasca project but on a much larger scale, yielding almost three times the amount of sweet crude. About 250,000 tons of tar sand are mined a day (above) and transported by a slurry pipeline to the extraction units. After separation, heated bitumen is fed into fluid cokers (opposite) where liquid and gaseous products are upgraded to yield synthetic crude oil and refinery gas. Employment at the job site peaked at more than 7,000 manual personnel, all living in a camp complex that had a full range of educational and entertainment facilities.

Syncrude, the second major project at the Athabasca tar sands, is all about scale. More than 50 million job hours and $1.8 billion were expended to create one of the largest material handling and refinery operations in the world. The Bechtel-built extraction facilities and upgrading plant produced some 125,000 barrels per day of sweet light synthetic crude oil from the oil-rich tar sands.

Some 300 miles northeast of Edmonton, the Syncrude project presented serious logistical problems, although nothing that Bechtel, with its extensive experience in the Canadian wilderness, hadn't conquered before.

For instance, four 350-ton Gofiner reactors had to be brought to the site from Italy. Because of the reactors' size, the trip across the ocean and then across Canada had to be planned with great precision. Indeed, the reactors were so heavy that a new bridge had to be constructed, and they could be moved only by rail in the dead of winter, when the roadbeds were frozen solid.

Another problem with the remote site was the lack of skilled workers and facilities. This, along with the complexity of the refinery, led Bechtel to use a preassembly strategy for construction.

Offsite preassembly of modules and components was not a new idea, but it had never been attempted before on this scale. Operations that were traditionally carried out at a job site were relocated to Edmonton. There, in an assembly-line operation, more than 2,500 workers at three sites built underground electrical duct banks, formwork, and concrete foundations; installed refractory lining in vessels; and prefabricated piping, sections of process plants, and pipeways, among many other tasks. Even sophisticated pieces of equipment and instrumentation were preassembled into modules and shipped north to the job site for final assembly.

Preassembly prevented material shortages, fostered better quality control, simplified labor requirements, and permitted controlled working conditions. It was not only cost-efficient, saving millions of dollars, but also time-efficient, helping bring the project to completion well ahead of schedule.

KING KHALID INTERNATIONAL AIRPORT
Saudi Arabia, 1974–83

The architecture of King Khalid International Airport is an example of Bechtel's ability to reflect the culture of a customer. Throughout, traditional Islamic motifs enrich this modern, highly functional complex.

On the geographic canvas of Saudi Arabia, big pictures are possible. One of them is King Khalid International Airport, built on an 87-square-mile tract (an area almost twice the size of San Francisco) 20 miles from the Saudi capital, Riyadh. The airport has five terminals—two international, two domestic, and one royal. Although the designs of most newer airports built around the world seem virtually interchangeable, this airport stands out. A modern gateway, it is unmistakably, and beautifully, Islamic in style. An adjoining community for 3,000 was built at the same time as the airport.

PALO VERDE NUCLEAR POWER PLANT
Arizona, 1976–88

Nuclear power plants, like fossil-fuel plants, need lots of water. So building a nuclear plant in the middle of the desert would seem as logical as planting tomatoes there. Yet the Palo Verde nuclear generating station was built in the Sonoran Desert, approximately 50 miles west of Phoenix, Arizona, where summer temperatures can reach 120 degrees Fahrenheit. Palo Verde is the largest nuclear plant in the United States, churning out 3,810 megawatts for the power-thirsty Southwest. It has plenty of water, thanks to an innovative idea: using reclaimed sewer water for cooling. Treated water is piped in from the Phoenix area, and Palo Verde's water plant produces up to 90 million gallons per day of clean cooling water. It leaves the plant only through evaporation.

There's plenty of cooling to do, too. Three identical units were built at the site. Bechtel engineers used a "cookie-cutter" approach, enabling workers to complete one reactor, then move to the next, where they would do the same work, reducing construction time and simplifying maintenance.

ALGERIA LIQUEFIED NATURAL GAS
Algeria, 1976–79

Bechtel's completion of Algeria's first large-scale liquefied natural gas (LNG) plant at the port of Arzew–El Djedid tripled the country's export capacity, enabling it to capitalize on its considerable reserves. The multibillion-dollar complex includes a six-train LNG plant and a 320-mile pipeline from the Hassi R'Mel gas fields in the Sahara south of the terminal facilities. At full capacity, the plant is capable of producing 370 billion cubic feet of LNG a year.

Gas produced in the Sahara is shipped via pipeline to Arzew, where it is liquefied by a cryogenic process. This reduces the volume of the gas to one six-hundredth of the space it would otherwise occupy. This is energy in a deep freeze, stored in immense tanks. The liquefied natural gas is then transported to market by huge cryogenic tankers.

JUBAIL
Saudi Arabia, 1976–Present

Jubail put the "giga" into gigaproject. For more than 20 years, Bechtel has been at work for the Royal Commission for Jubail and Yanbu, helping build a modern metroplex at the site of what was once a fishing village on the Arabian Gulf. Now the 360-square-mile area includes a complete industrial and residential infrastructure, 16 primary industries, and a planned community expected to house as many as 370,000 people.

Jubail traces its origins as much to the imaginations of the Saudi royal family and Steve Bechtel Sr. as to Saudi Arabia's rich deposits of oil and gas. Steve Sr. saw gas being flared off each year and wondered what could be done with it. At the same time, the king was looking for an opportunity to move his country beyond a natural resource economy to a more diversified one. The initial concept was to use natural gas as a power source, then build

For more than 20 years, Bechtel has worked closely with the Royal Commission for Jubail and Yanbu (opposite), to transform a sleepy fishing port into an industrial powerhouse (above). Jubail has become home to the world's largest producers of petrochemicals.

We used to think of the Panama Canal as a big project,
but the Canal is one-tenth the size of Jubail.

STEPHEN D. BECHTEL SR.

Many of Jubail's industrial and utility components were built elsewhere, shipped to the new port, and then moved into place to be installed. This modular concept was necessary because of the scope of the work and the speed needed to combine both residential and industrial development. That made for challenging problems—it's not every day that a factory arrives on a truck.

industries that were either heavy power users, natural gas-based, or both. A planned city would provide housing and residential amenities for workers and their families. Another, smaller effort was undertaken at Yanbu on the Red Sea coast.

But as the original $8 billion program started to take shape in 1977, the Saudis began thinking on an even bigger scale, making Jubail the biggest civil engineering project on Earth. Even the site preparation work was impressive: The mean elevation of the entire area had to be raised about seven feet.

The first step was to build infrastructure, starting with a power supply. A network of highways, an airport, and housing were designed and constructed from scratch. A telecommunications system was built, as well as a seawater cooling system that could generate a water supply equal to two-thirds the annual flow of the Tigris and Euphrates rivers. Bechtel would help with an industrial port and others would build a commercial one.

The first primary industry, the steel mill, came on-line in 1982. It would be followed by refineries, petrochemical plants, and more than a dozen industrial plants making everything from fertilizer and plastics to a range of petrochemicals and petroleum products. This vast and constantly expanding industrial base has made Jubail a major force in world petrochemical markets. By the time a 15-year development review was completed in 1992, Jubail's 70,000 residents had 22 schools, 14 shopping centers, and more than 300 commercial businesses.

But the Saudis were interested in getting more than factories and offices. They wanted to build the capabilities of their own people, too, through a technology transfer and training program.

Jubail is still a work in progress, and Bechtel will continue to be part of it. Recognizing 20 years of successful management, the commission recently extended its agreement with Bechtel for another decade.

Jubail is not just an industrial park; it is also a residential development. The coastal community (opposite) was built in tandem with the industrial facilities so that enough housing would be available for workers as the new plants came on-line. Other structures include state-of-the-art government buildings, such as the Royal Commission administrative offices (above).

THREE MILE ISLAND CLEANUP
Pennsylvania, 1979–90

Because they had built dozens of nuclear power plants around the world, Bechtel engineers knew as much as anyone about the workings of reactors. But they didn't know the extent of the damage created on March 28, 1979, when a valve failure and operator error at the Three Mile Island nuclear power plant's Unit 2 exposed the fuel core, causing it to overheat and begin to melt. Nobody knew.

Brought in by the reactor's owner, General Public Utilities Corporation, Bechtel's Nuclear Fuels Operations group (with help from Bechtel Power) would have to find the damage and figure out a way to fix it, without being able to go into the damaged reactor vessel. (Three Mile Island was designed and built by other companies.) Although Bechtel's engineers can and do simulate plant accidents in their research, one of them described the project as being like "somebody handing you a 1,000-piece jigsaw puzzle without the picture on the box."

Acquiring data about the core condition proved to be one of the most important and challenging tasks. It would be three years before a camera could be lowered into the reactor.

Working from a control room in the turbine building (above), technicians manipulated specially designed robots to decontaminate the basement of the reactor building. Once inside the containment building (opposite), engineers and technicians were able to gather data, maintain plant controls and equipment, and install television cameras to monitor the cleanup.

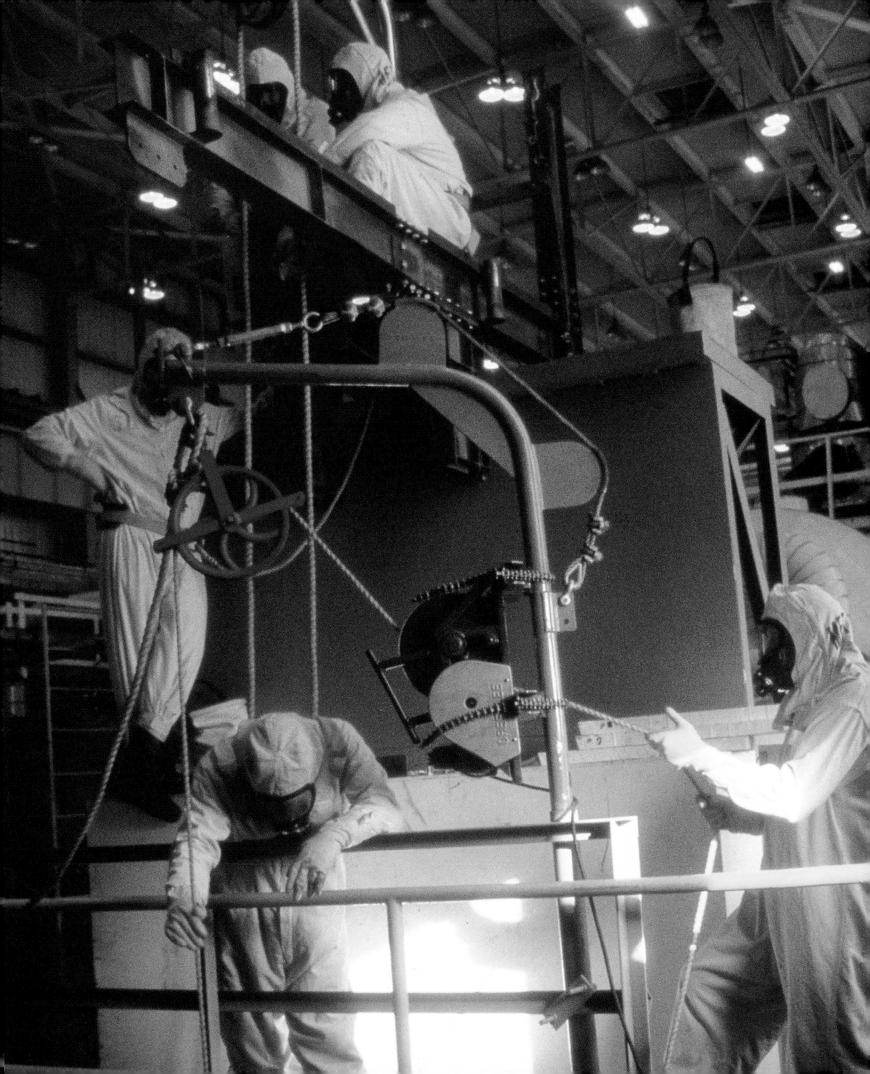

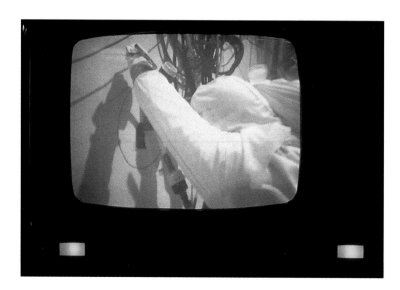

The camera would show that the top five feet of the 12-foot-high fuel assemblies—177 of them—had shattered from high temperatures followed by rapid cooling. A five-foot mass of once-molten rubble littered the floor. Radioactive water filled the basements of the containment and auxiliary buildings.

Working deliberately, imaginatively, and with great attention to safety, Bechtel engineers developed the cleanup plan, as well as many of the tools necessary to do the job. A number of robots were designed to collect information on radiation and carry out the task of decontamination. Fuel removal, which began in 1985, was completed in 1989.

Workers clad in protective suits, seen on an internal monitor, vacuumed, drilled, and jackhammered the damaged core until pieces were small enough to fit into special holding vessels.

Despite all the high-technology techniques, nature delayed the cleanup in early 1986. The river water that had leaked into the containment building was filled with microorganisms (bacteria, fungi, yeasts, algae) that had lain dormant for years. Heat inside the reactor vessel caused these microorganisms to start multiplying, and they began to thrive once the cleanup had begun, clouding the water beyond visibility. They were eventually destroyed by hydrogen peroxide.

When its work was done, Bechtel left Three Mile Island a stable and secure facility that can be maintained for an extended period of time.

OK TEDI GOLD AND COPPER MINE
Papua New Guinea, 1981–84

A mountain of gold would seem hard to miss. But when the mountain is located in Papua New Guinea, such an oversight becomes understandable. The location, Mt. Fubilan, is more than a mile above sea level and protected by dense jungle. The inhabitants had no contact with outsiders until 1963.

But with an estimated 40 million tons of gold ore sitting on a trove of 400 million tons of copper ore, Mt. Fubilan was too rich to pass up. So the Ok Tedi project took on the jungle. Bechtel and Morrison-Knudsen International had the assignment to design and build a gold-processing plant. Then, a month before the scheduled start date, the company was asked to complete an 80-mile road to the construction site.

Bechtel was familiar with the region and had extensive experience with remote projects in all types of terrain. But Mt.

Ok Tedi presented engineers with a challenging combination of geology and geography: gold sitting on top of copper and all of it sitting on top of a cloud-shrouded mountain (above) deep in an equatorial rain forest. Bechtel, with Morrison-Knudsen International, designed the gold-processing plant (opposite), and then transported the component parts to the remote site piece by piece.

Fubilan protected its treasure tenaciously. Hundreds of thousands of tons of equipment, material, and supplies had to be barged 800 miles from Port Moresby, across the Gulf of Papua, and up the Fly River. At the mine site, rain fell an average of 339 days a year, making it very insect-friendly. Trucks and heavy equipment sank in the mud. Gravel disappeared under standing water. Then, suddenly, the rain stopped, taking with it the most efficient means of transportation, the Fly River. C-130 Hercules aircraft replaced barges, transporting essential supplies, food, and fuel to keep the project going until the drought broke.

The Fly River was the most efficient transportation route through the dense rain forest until an unexpected drought made it impassable, and thousands of tons of supplies had to be transported by air.

The rains returned only to cause another near-catastrophe. Sixty-five million cubic yards of hillside slipped into the Ok Ma Valley, where engineers had planned to build a tailings dam. An alternative short-term plan was developed to complete the mine on schedule, and the first gold was produced in May 1984.

NEW ZEALAND GAS-TO-GASOLINE

New Zealand, 1982–85

Hydrocarbons don't always present themselves in the most desirable form for energy use. For instance, New Zealand has plenty of natural gas from its Maui field in the Tasman Sea just 20 miles off the coast of the North Island. But it lacked significant petroleum reserves, a point brought home powerfully during the energy shortages of the 1970s. Mobil came up with a solution: converting natural gas to methanol and then, using a new catalytic process, converting the methanol into high-octane, low-lead gasoline. Bechtel came up with the plant, the first of its kind.

Much of the equipment had to be sourced elsewhere. From Japan came tremendous modules, some weighing as much as 646 tons. The job also required Bechtel's project finance capability as well as its logistical expertise to handle export-import financing and to coordinate work with nearly 20 different New Zealand federal and local agencies.

LOS ANGELES SUMMER OLYMPICS

California, 1983–84

There is joy, of course, for the organizing committee that wins the right to hold the Olympic Games. And then comes the day after, when somebody realizes that dozens of facilities are needed for the event, and must be completed under great time pressure. That's the position the city of Los Angeles found itself in when it called on Bechtel with a last-minute plea to help finish the planning, design, and construction of the 1984 Summer Olympic facilities.

The 40-person Bechtel team, led by Ed Keen, crossed the finish line in time, coordinating work completion in 60 separate locations spread over three counties, many of which were inaccessible until the last minute. Most of the work had to be accomplished within six weeks, and, because the Games' organizers planned to use existing facilities, nine sites had only a two-week construction window. In total, sites had to be made ready for 23 sports. Work included construction of a new swimming pool and velodrome, 14 training centers, 3 Olympic Villages, and 19 support facilities.

"There's no better feeling than pulling off the nearly impossible and seeing your efforts make a difference," said Bob Polvi, president of the Bechtel unit that completed preparations for the Los Angeles Summer Olympics in record time.

The job was right up Bechtel's alley, given the company's demonstrated ability to handle the cost and scheduling of megaprojects and to manage a large number of subcontractors simultaneously.

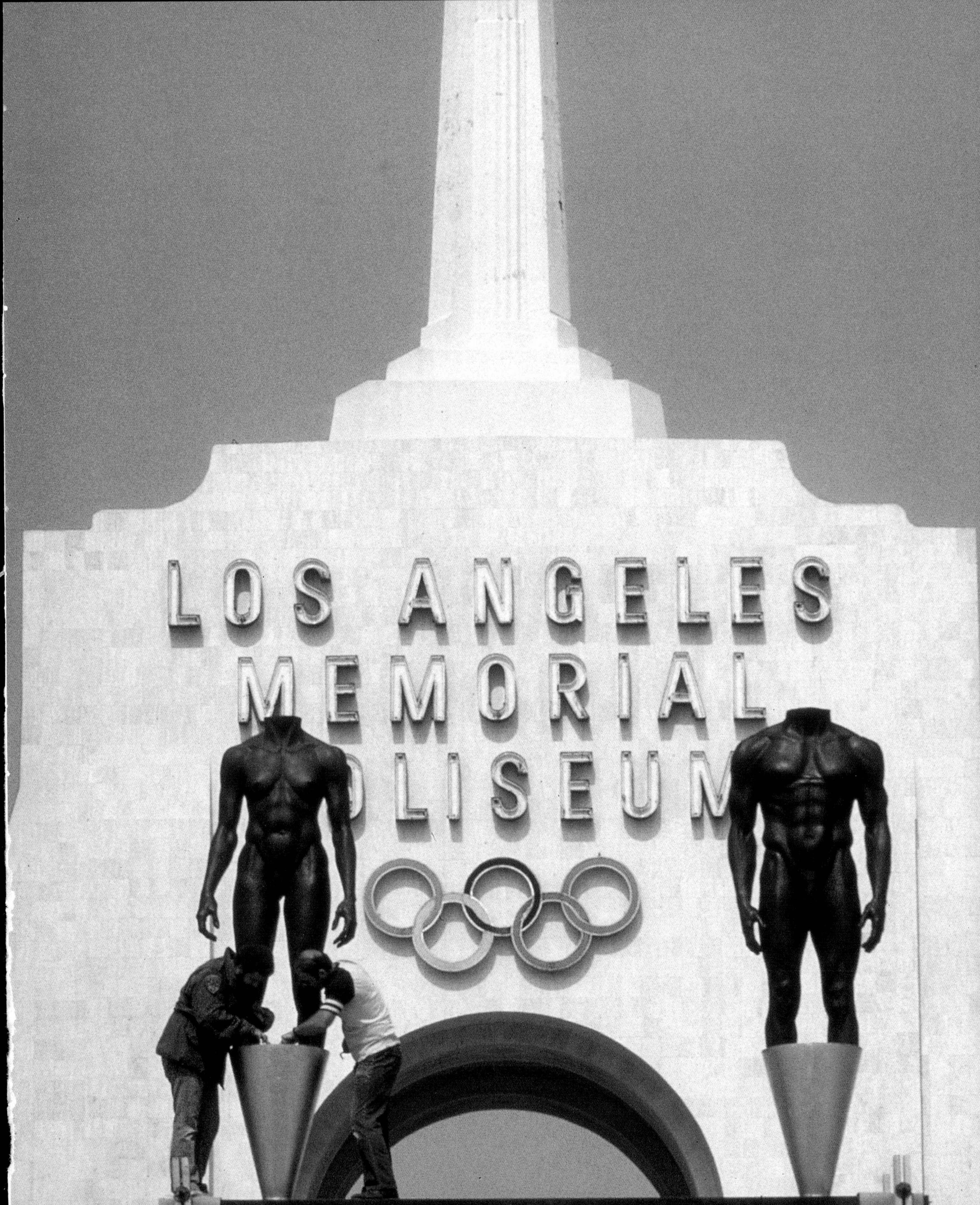
LOS ANGELES
MEMORIAL

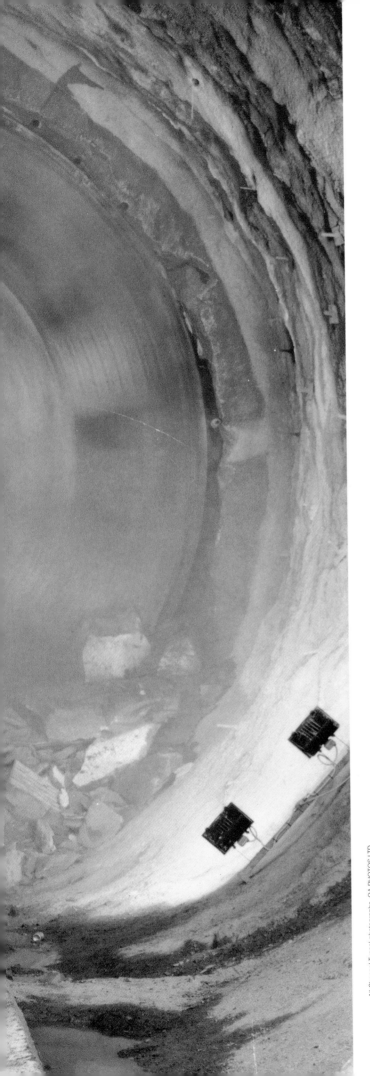

CHANNEL TUNNEL
England-France, 1986–94

The search for the best means to cross the English Channel has challenged great minds—even Napoleon's—for the last two centuries. The first practical plan was tested in 1867 with a giant tunnel-boring machine, but military considerations brought the effort to a halt. In the mid-1970s, another attempt was thwarted by financial concerns.

The much-dreamed-of tunnel and transportation system would eventually provide a 32-mile undersea leg between Folkestone in England and Calais in France. Specially designed trains would carry passengers, cars, trucks, and buses below the English Channel. The project began in earnest in 1986. As work progressed, the owner, Eurotunnel, and the Anglo-French consortium responsible for design and construction, TransManche Link, were plagued by severe cost, schedule, and safety problems, and by 1990 there were fears that the project would never be finished. Bechtel's involvement began in 1987, mainly in a key management role. But when it became

A monstrous 400-ton boring machine took a bite out of the limestone beneath the English Channel (opposite). Each terminal (above) was a major civil engineering project in itself, housing maintenance and service depots, a power substation, and an operations control center.

All Channel Tunnel photographs, QA PHOTOS LTD.

203

clear that the project was in danger, Eurotunnel called upon the resources of Bechtel to play a larger role. Bechtel's experts, led by John Neerhout Jr. as project chief executive, were given full responsibility to get the project back on track—and that's exactly what they did.

As a tunnel-boring machine made its way along the route, guided by lasers and satellite navigation, it left behind curved concrete tunnel liners. Crossing alleys were dug out with smaller equipment.

At the peak of construction, eleven giant tunnel-boring machines, each costing up to $20 million and weighing as much as 1,200 tons, chewed their way through subaqueous chalk; it took 150 electric and diesel locomotives to haul away the "muck." The length of the tunnel required scrupulous safety measures. Every 400 yards, there are cross passages for ventilation, maintenance, and emergency evacuation on the three tubes (two single-track railroad tunnels flank a narrower service tunnel). Also, two huge 400-kilowatt fans are housed atop each access shaft, from which fresh air flows into the running tunnels. But the greatest concern is fire. With up to 20,000 people in the tunnel at one time, some more than 10 miles from open air, the strictest safeguards had to be applied. Among those protections are electrical equipment chambers with fireproof doors, fire extinguishing systems, shuttle cars with ultraviolet and opacity smoke detectors, foam and Halon gas extinguishing systems, and doors providing 30-minute fire protection. The tunnel can be evacuated within 90 minutes, if necessary.

The privately financed project, which ultimately cost $14.7 billion, has made possible 500 undersea train trips a day at speeds of up to 99 miles per hour. For travelers, the Channel Tunnel has been a boon. A trip from central Paris to central London can now be made in just three hours by train. This time will be cut to two hours and 20 minutes on completion of the Channel Tunnel Rail Link projected for 2007, which Bechtel is now designing and building. To be built in two phases, the link will run from Folkestone on the coast to London's St. Pancras Station.

About 48 miles [of tunnel] were dug in 1990, and that's a world record.

ROGER PICARD, BECHTEL SENIOR VICE PRESIDENT, MANAGING DIRECTOR OF CONSTRUCTION

TRANS-TURKISH MOTORWAY (ANKARA-GEREDE)
Turkey, 1986–98

Bechtel's work on the Trans-Turkish Motorway, with its local partner Enka, is a perfect demonstration of the company's global reach paired with its unique ability to deliver everything, from crushed rock to sophisticated financing, to a customer—in this case the Turkish government.

Enka-Bechtel built two sections of the 875-mile high-speed motorway. The highway system is critical to Turkey, linking Istanbul, Turkey's European port, and Ankara, its Asian depot.

Building the motorway was a challenge. In some places, the road reached heights of 5,000 feet; altogether more than 196 million cubic yards of earth and rock were excavated. The first Enka-Bechtel section, a six-lane, 70-mile segment, connects Ankara, the capital, with the city of Gerede; the second is an eight-lane, 70-mile beltway that circles Ankara.

At peak, the $1.4 billion effort employed nearly 6,000 Turkish workers—one of Bechtel's largest international direct-hire jobs—and raised construction skill levels in Turkey. In some cases, workers who had arrived at the site not knowing how to drive a car left as expert operators of multimillion-dollar, 250-ton hydraulic cranes.

Often running alongside ancient oxcart paths (below), Turkey's new network of motorways is nevertheless modern in scale. The high-speed toll road is an integral part of a transcontinental highway system linking Europe with Asia. The completed motorway will cut travel time between the commercial centers of Istanbul (above) and Ankara by up to 50 percent.

BOSTON CENTRAL ARTERY/TUNNEL
Massachusetts, 1986–Present

Boston is more gridlock than grid, and its drivers are legendarily impatient. Who can blame them? They have to drive on I-93, the Central Artery that divides the neighborhoods near Boston's waterfront from the rest of the city. To Beantowners, I-93 is more like a Central Nightmare.

Help is on the way. Under the joint management of Bechtel and Parsons Brinckerhoff, the Central Artery/Tunnel project, the biggest civil engineering undertaking in the country, will reconnect Boston and delight its beleaguered motorists. The 7.5-mile elevated portion of I-93 will be replaced with an 8-lane tunnel through downtown Boston and a 10-lane cable-stayed bridge across the Charles River. The project also includes the recently completed Ted Williams Tunnel, an extension of I-90 under Boston Harbor to and from Logan International Airport. The tough part of the project? Building a new 8- to 10-lane highway beneath the old one without closing the old road, and at the same time trying not to disrupt life in Boston's neighborhoods. For instance, how do you remove 13 million cubic yards of fill without making a mess? It takes precision engineering and precision community relations.

Boston Central Artery is designed to relieve terminally congested I-93, a six-lane elevated highway that has cut downtown Boston off from its waterfront for 40 years. "The Big Dig," as Boston residents refer to the Central Artery/Tunnel project, will reunite the downtown area with the city's waterfront.

Andy Ryan

On most construction jobs, you keep people away from the work. On this job, our most important task is keeping work away from the people.

MEL MIRSKY, BECHTEL PROGRAM MANAGER

SAN FRANCISCO MUSEUM OF MODERN ART
California, 1989–94

With its distinctive Modernist style, SFMOMA *is a new architectural landmark for the city (above). In the center of the museum, a skylit atrium rises 145 feet to the full height of the building (opposite). Inside, the galleries are flooded with natural light.*

The best new museum building America has seen in years," raved *Time* magazine art critic Robert Hughes on viewing the San Francisco Museum of Modern Art. Had he seen how Bechtel's project management had saved millions in costs while executing Swiss architect Mario Botta's design, Hughes might have been even more impressed. The 225,000-square-foot building, located in San Francisco's South-of-Market area, features five floors of exhibit space, each of which is stepped back at the roof line to maximize the natural light flowing into the galleries. The focal point is a five-story atrium topped by a skylit cylinder built of alternating bands of white and black stone.

Bechtel, a South-of-Market neighbor, agreed to provide project management services at cost, an initial saving of $1.6 million. Then the project team undertook budget reviews that cut projected construction costs of $65 million by more than $12 million. Most of the savings came in the form of "value engineering," the ability to deliver more for the money through imagination and innovation. Instead of using a concrete structural frame, for instance, Bechtel suggested a structural steel frame that saved 10 to 15 percent on steel production and installation costs. Instead of laying brick by hand, architectural precast panels with brick facings were used, speeding up the brickwork by some 50 percent. And change orders, the bane of all construction projects, were all but eliminated. The result is a spectacular and cost-effective addition to the Bay Area fine arts scene.

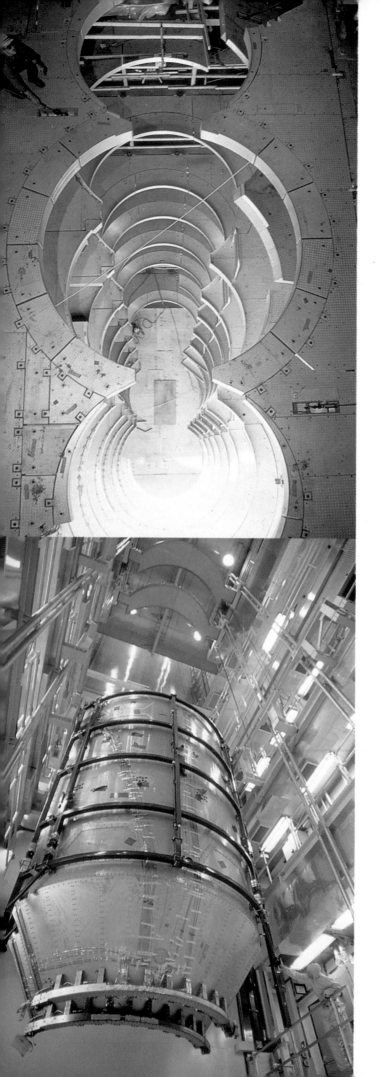

SPACE LAUNCH COMPLEX 40
Florida, 1990–92

It is 23 stories tall and weighs 11.4 million pounds—and it has to move. The mobile service tower of Launch Complex 40 at Cape Canaveral is the most massive self-propelled building in the world. A sister complex at the Cape had taken five years to complete. Bechtel was expected to complete this one in 24 months, in time to launch the *Mars Observer* probe in the fall of 1992, when Earth and the Red Planet would be ideally aligned. The launch window would not open again for five years. Planet alignment, noted Bechtel Executive Vice President Bill Friend, is not something you can renegotiate if you fall behind schedule. This new job also included demolishing and rebuilding most of the other support facilities at LC-40, where the *Observer*'s Titan III launch vehicle would blast off. There was no time to waste.

During the course of work, more than 150 change orders would essentially double the scope of work on the mobile service tower. Bechtel pulled out all the stops, sending design engineers from San Francisco to follow their work modules to the Cape from design through construction through start-up engineering. At one point, 1,000 workers swarmed all over the complex, laboring in two 10-hour shifts, seven days a week. In the end, Bechtel met its schedule, and so did the *Mars Observer* probe, lifting off on September 25, 1992.

The mobile service tower seals the payload inside a class-100,000 clean room (above and below, left), a virtually dust-free environment shielded against electromagnetic interference and electronic eavesdropping. Inside this space, technicians make final connections and adjustments, a process that may take two months or more.

*The Mars probe captured the imaginations of
everyone working on Launch Complex 40, and
I think that's part of the reason we were able to
complete a nearly impossible project successfully.
It was the sense we were all involved in something very grand and important.*

WAYNE BORGER, BECHTEL'S SITE CONSTRUCTION MANAGER

HONG KONG
AIRPORT CORE PROGRAMME
Hong Kong, 1990–98

Both photographs, Andy Ryan

As would befit the world's largest airport, the creation of the 3,100-acre site for the Hong Kong International Airport was a massive undertaking (above). The majority of the world's high-capacity dredging fleet was used to redistribute approximately 450 million cubic yards of rock and soil, and both Chek Lap Kok and a small neighboring island

were leveled. The airport includes a 5.5 million-square-foot terminal building with eight levels and 38 fixed gates, each able to accommodate the world's largest passenger aircraft. From the airport, the six-lane North Lantau Expressway (opposite) and a high-speed airport railway carry travelers to the urban areas of Hong Kong.

When the British handed over sovereignty of Hong Kong to the People's Republic of China in June 1997, the transfer included the nearly completed Hong Kong International Airport, due to open in 1998. Hong Kong's harried travelers would finally find relief from the busy Kai Tak Airport, where jets landed between neighboring high-rises. Kai Tak, with an official capacity of 24 million passengers, handled nearly 30 million in 1997; the new airport accommodates 35 million before a second runway and future expansions are completed.

Since it was impossible to build an airport in an island city where every square inch of land is already stacked with high-rise buildings, the only solution was to grind two small, remote islands down to sea level, build an artificial island, and then lay a state-of-the-art airport over that. The airport island would have to be connected to the mainland by highways and

mass transit. Hong Kong International Airport is the cornerstone of a remarkable 10-part, US$20 billion infrastructure development known as the Hong Kong Airport Core Programme, designed to alleviate some of the congestion associated with Hong Kong's continuous growth as the center of business for greater Asia, and the commercial gateway to China.

With so much on its plate, Hong Kong's New Airport Projects Coordination Office (NAPCO) brought in Bechtel as a program management consultant. With Bechtel's assistance, NAPCO took a rigorous approach and completed the vast undertaking on time and with budget savings in excess of a billion dollars (U.S.).

The NAPCO roster is an urban planner's dream list. It includes the 3,100-acre airport and a 7,086-foot suspension bridge. The bridge connects to the 7.7-mile North Lantau Expressway, one section of a new 21-mile transportation corridor. The project also includes a third tunnel under Victoria Harbour, a 21-mile mass transit rail running from Hong Kong to the new airport, two reclamation efforts on Kowloon and Hong Kong Island to win back some 875 acres from the water, and the Tung Chung development next to the airport, initially with housing for 20,000 people.

The Lantau Link (above), which joins Lantau Island to the mainland, comprises two bridges: the cable-stayed Kap Shui Mun, with its distinctive "H"-shaped towers, and, in the distance, the Tsing Ma suspension bridge. Spanning a mile and a third, the double-decked Tsing Ma Bridge is the longest bridge ever to carry both rail and road traffic.

Both bridges required designs that could withstand typhoons, which regularly sweep in from the South China Sea with winds of more than 125 miles per hour. The final stage of the expressway is a 1.2-mile tunnel under Victoria Harbour with three lanes in each direction, connecting Kowloon with Hong Kong Island (opposite).

Andy Ryan

Iraqi demolition teams had rigged Kuwait's oil wells for maximum destruction, directing blasts downward by piling sandbags on top of charges. Firefighters faced hundreds of blazing wellheads. Even the day after a fire was killed, the earth was still hot enough to boil water.

KUWAIT OIL FIRES
Kuwait, 1991

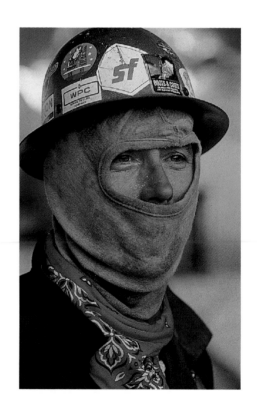

A giant blanket of smoke hung over most of Kuwait; people had to navigate the streets by flashlight during what should have been broad daylight. The magnitude of the devastation wrought by the Iraqi attack on Kuwait's oil fields was unimaginable: 650 wells ablaze and much of the area a minefield of unexploded ordnance. Experts predicted that the Kuwait fires would burn for five years, with potentially catastrophic results. Bechtel defied the odds, and its team of experts snuffed out the flames in only eight months.

The heat was unbearable; the roar of fires being fed crude under high pressure was like that of a 747 at full throttle. Huge lakes of burning oil covered the ground, flooded the roads, and hid some of the mines, making it impossible to get near many of the fires. The wells were spewing viscous crude onto the desert floor, where it pooled into vast swamps of thick black muck.

The quenching of Kuwait's oil fires is first the story of

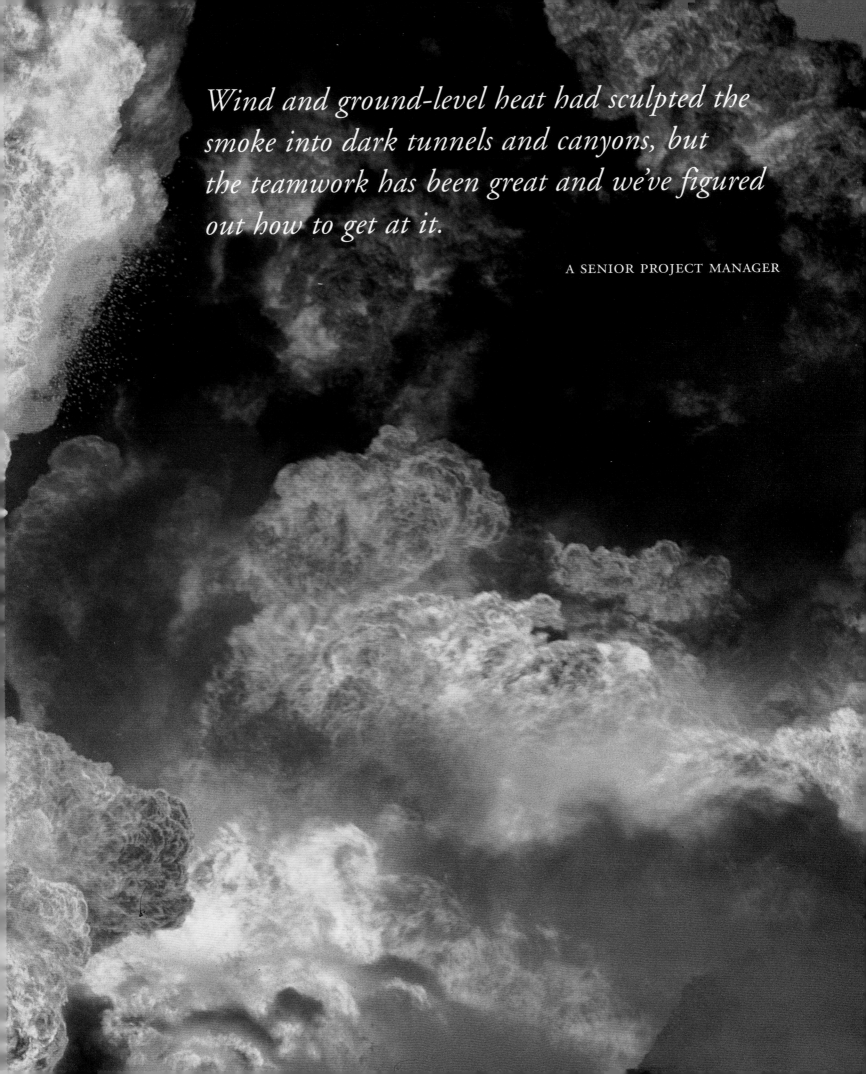

Wind and ground-level heat had sculpted the smoke into dark tunnels and canyons, but the teamwork has been great and we've figured out how to get at it.

A SENIOR PROJECT MANAGER

a massive, Bechtel-managed logistics effort, perhaps the biggest in the company's history. In less than six months, a city of management staff, engineers, technicians, and equipment operators was set up in the desert. Using the port of Jebel Ali in Dubai as a staging area, Bechtel marshaled 125,000 tons of equipment and supplies, including some 4,000 pieces of equipment ranging from bulldozers to ambulances. The enormity of the disaster and the high-stakes competition to complete the cleanup as quickly as possible led to a number of technological breakthroughs.

But to begin with, no water was available to tame the fires. And, above all, the firefighters needed water: 20 million gallons per day. Within two weeks of Kuwait's liberation, Bechtel had decided to use the pipelines that normally flowed crude to the ports and reverse them to pump seawater back to the fires. Gushing oil was dammed to the side; a road to the wellhead was graded over the burning ground; the

To gain access to the burning wellheads, crews first had to clear the area of live munitions. Then, bulldozers prepared a road to the wellheads by covering the burning sand with compact landfill.

assault was ready to begin. The goal was to beat back the fire, stop the gushing oil by using a boom to get a pipe nozzle in place, and pump in a mud-like material to seal off the well.

At the same time, a recovery team puzzled over how to handle the lakes and pits of oil. A field engineer from Bechtel's London office directed development of an experimental recovery pond with skimming devices and a rudimentary system to clear contaminants. The effort squeezed 10,000 barrels of oil a day from the saturated sands.

On November 6, 1991, Sheikh Jaber Al Ahmed Al Sabah, the emir of Kuwait, closed a valve in the Burgan field, snuffing the final fire.

TOYOTA TEST TRACK

Arizona, 1991–93

Toyota's Arizona proving ground is one of the largest test tracks of its type, encompassing 20 square miles—all of it fenced off to prevent industrial spying. The facility, designed and built by Bechtel with Japan's Taisei Corp. for Toyota Technical Center, is a high-speed performance, endurance, and reliability evaluation complex. Its star attraction is a 10-mile, three-lane, velvety smooth high-speed oval track, built so accurately that a driver can take a corner at 160 miles per hour without hanging on to the steering wheel. Other features include a cornering course to test handling and stability, a hill section, and a material exposure area to evaluate weather resistance of parts. Because it is in the Arizona desert, the track is also used for high-temperature evaluations. The heat made for interesting construction twists. Concrete had to be poured at night, with crews using ice instead of water to keep it cool enough to set properly. The site also has some reality-based road terrain, carefully designed to match the cracked, pothole-ridden road conditions drivers are accustomed to seeing. It's some of the best bad roadway ever built.

Predawn concrete pours are a must in the desert heat (opposite). The curves on the test oval angle 29 degrees at the outer edge, requiring a special computer-operated paving machine (above).

AIRCRAFT TEST CENTER
Florida, 1991–97

The coldest weather in Florida is always in the same spot: the McKinley Climatic Laboratory at Eglin Air Force Base near Pensacola. Since 1947, the lab has taken virtually all the aircraft the Air Force puts into service, including the Stealth fighter, and run each through a grueling test of temperature extremes and storms to make sure it can operate under adverse conditions.

It would stand to reason that the building has taken a lot of punishment. Temperatures in it during tests range from minus 65 degrees to 165 degrees Fahrenheit. McKinley was built to perform for 30 years; it was already pushing 50 when Bechtel was engaged by the U.S. Army Corps of Engineers to plan a major overhaul. McKinley was showing its age; thousands of freeze-thaw cycles had displaced the floor some eight inches in spots, one of several threats to the vapor barrier critical to McKinley's operations. The project included revamping the two largest test chambers, one of them big enough to hold the monster-size C-5A Galaxy cargo

Aircraft are typically tested for three to six months at McKinley Climatic Laboratory, where they are subjected to everything from extreme heat and cold to sand and ice storms.

transport. To bring the facility up to date, Bechtel designed new air-handling systems to accommodate the mightiest jet engines at full power. Bechtel had to make sure the engines, operating indoors, got a new supply of air as quickly as they used up the old. Without the fresh air, the engines could literally suck in the walls. A year after project completion, the walls of McKinley have not budged.

ATHENS METRO
Greece, 1991–Present

The Acropolis keeps watch over the Athens Metro project. Omonia Square (opposite) was perhaps the trickiest site; it required construction of a new tunnel and station below the existing subway line, along with renovation of the concourse above the current station—all in the middle of the city's busiest intersection.

The Athens Metro project was designed to relieve the historic city of some decidedly modern problems. Its traffic-clogged streets have led to air pollution—*nefos*, in Greek—severe enough to cause harm to historic buildings, including the Parthenon.

Urban projects generally require a community outreach program to inform locals of the disruptions that might occur. In Athens, the company faced a different kind of public: Skeptical Athenians were not convinced that the job would ever be done. A subway had been announced dozens of times before, and the prospect of another one just made them chuckle.

But with Bechtel supplying management and engineering services to Attiko Metro, a private company established by the Greek government to manage construction, the chuckles soon turned to cheers. Eleven miles of tunnel and 21 stations are being built for two new lines. The lines form an "X" crossing in Syntagma Square, in front of the Parliament Building. The Greek government was so pleased with the work that it announced a $1.2 billion expansion.

Of course, in a city as old as Athens, a subway project inevitably becomes an archaeological dig. And dig Bechtel did, participating in the largest archaeological excavation in the city's history. (This was a dream job for Bill Stead, the job's first top manager, who happens to be a trained archaeologist.) Five of the 21 new Metro stations fell within the boundaries of ancient Athens and have yielded artifacts as diverse as a bathhouse, aqueducts, city walls, and roads—the works of some classical engineer-constructor, no doubt.

MOTOROLA FACILITY
China, 1992–Present

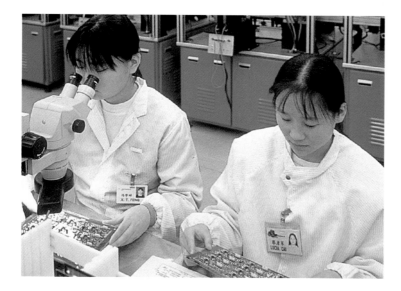

One of the more remarkable aspects of wireless tele-communications is that it allows countries with poor telecommunications infrastructures to vault directly into the digital age, bypassing more developed nations in the process. That was China's goal, and in Motorola the Chinese found the partner to get them there.

As Motorola knew, it couldn't build high-technology products in China unless it had high-quality facilities. That is why the company asked a Bechtel joint venture, China American International Engineering, Inc. (CAIE), to design and build its manufacturing facilities in Tianjin. At this writing, 592,000 square feet are used for production of pagers, semiconductors, and cellular telephones as well as for office space.

The Motorola China operations are world-class, manufacturing products that meet the company's Six Sigma quality standards, a statistical measure that allows no process yield defects greater than 3.4 for every million parts of production.

SOLAR II
California, 1993–98

Mirrors on the desert floor (opposite) were aimed at a receiver tower 300 feet tall (below). A tank at the tower's base stored solar energy for several days without significant heat loss, allowing a turbine to generate electricity during a storm and even at night.

I n the Mojave Desert, Bechtel put more than two decades of solar energy experience to work as part of a government-sponsored consortium to explore an advanced solar thermal technology called "molten salt heat transfer."

Solar II is located at the site of an earlier experiment. Nearly 2,000 heliostats (mirrors) direct the desert sun to a 300-foot tower. A combination of sodium and potassium nitrate—like the chemicals in garden fertilizers—is pumped from a holding tank on the ground to the top, where the highly concentrated sunlight heats the molten salt to 1,050 degrees Fahrenheit. The salt then flows down the tower to a second, insulated holding tank. From here, it can be pumped to a heat exchanger that produces superheated steam to drive a turbine generator, or it can be stored and its thermal energy used later. Located near Barstow, California, the plant can produce 10 megawatts of electricity, enough power to supply 10,000 homes. The program participants hope to have demonstrated that power plants generating 10 or 20 times as much energy will be feasible using this technology.

Tim Bieber, The Image Bank

NAVY CLEAN II

1993–Present

H urry up and wait" characterizes one of the more frustrating modes of military operations. But in the case of environmental cleanup at U.S. Navy and Marine Corps bases, there was no time to lose. Many sites were being turned over to the public for civilian use, and every day wasted increased the chances of further environmental damage.

Since 1993, Bechtel has been part of the U.S. Navy's CLEAN II (Comprehensive Long-Term Environmental Action Navy) team. In the Southwest, as project managers, Bechtel has handled investigations at 42 installations that contained hundreds of potential problem sites.

The Navy's problems were mainly administrative. Previously, site cleanups had been subject to an endless loop of submissions between contractors, subcontractors, and military and civilian review organizations.

Bechtel's management approach collapsed the process into a team-driven, action-oriented program, reducing the cycle time per project by more than two-thirds. The company developed a site management plan, bringing together subcontractors and their Navy counterparts to set goals and foster clear communication.

In similar work for the Navy that is not part of the CLEAN program, Bechtel is directly involved in major environmental cleanup at installations in Florida, Georgia, and the Carolinas. In the Southeast, Bechtel is responsible for actual cleanup at 16 U.S. Navy and Marine Corps sites. Bechtel teams do everything from keeping contaminants at bay in environmentally sensitive areas such as Key West to cleaning up petroleum-contaminated groundwater at the Athens Naval Supply Corps School in Georgia.

Bechtel engineers created "virtual vistas" of the future Dubai International Airport. The images are readily changed by 3D-CADD (computer-aided design and drafting) programs and easily "experienced" with virtual reality technology.

VIRTUAL REALITY
Dubai International Airport, 1996

While most of the development in virtual reality computing has been conceived by entertainment media, Bechtel is exploring its practical applications. From designing therapy centers that treat brain tumors to building airports, virtual reality models from Bechtel engineers are changing the future.

At present, Bechtel is using virtual reality to change the future for Dubai, one of the United Arab Emirates. Dubai, the main port and commercial center of the region, relies on trade and tourism rather than oil for its economic growth. Its current airport serves a hundred destinations—more than any other Arab nation. An airport center with every conceivable facility will transform Dubai's airport into a major hub of the Middle East.

Bechtel's engineering team took Dubai's Civil Aviation Department executives on a virtual reality tour of their new airport, with a half-mile-long, state-of-the-art multistory concourse, a new control tower, 28 new arrival-departure gates, a full-service conference center, a hotel, an entertainment center, and the world's largest duty-free shop. And not a brick had been laid. Thanks to computer modeling, changes to this nonexistent facility can be made right on the spot.

The use of virtual reality on the Dubai project has also served as a pilot for many other programs, including documentation systems that will provide detailed archives, among other uses.